Recent Titles in This Series

- 41 **Hal L. Smith,** Monotone dynamical systems: an introduction to the theory of competitive and cooperative systems, 1995
- 40.1 **Daniel Gorenstein, Richard Lyons, and Ronald Solomon,** The classification of the finite simple groups, 1994
- 39 **Sigurdur Helgason,** Geometric analysis on symmetric spaces, 1993
- 38 **Guy David and Stephen Semmes,** Analysis of and on uniformly rectifiable sets, 1993
- 37 **Leonard Lewin, Editor,** Structural properties of polylogarithms, 1991
- 36 **John B. Conway,** The theory of subnormal operators, 1991
- 35 **Shreeram S. Abhyankar,** Algebraic geometry for scientists and engineers, 1990
- 34 **Victor Isakov,** Inverse source problems, 1990
- 33 **Vladimir G. Berkovich,** Spectral theory and analytic geometry over non-Archimedean fields, 1990
- 32 **Howard Jacobowitz,** An introduction to CR structures, 1990
- 31 **Paul J. Sally, Jr. and David A. Vogan, Jr., Editors,** Representation theory and harmonic analysis on semisimple Lie groups, 1989
- 30 **Thomas W. Cusick and Mary E. Flahive,** The Markoff and Lagrange spectra, 1989
- 29 **Alan L. T. Paterson,** Amenability, 1988
- 28 **Richard Beals, Percy Deift, and Carlos Tomei,** Direct and inverse scattering on the line, 1988
- 27 **Nathan J. Fine,** Basic hypergeometric series and applications, 1988
- 26 **Hari Bercovici,** Operator theory and arithmetic in H^∞, 1988
- 25 **Jack K. Hale,** Asymptotic behavior of dissipative systems, 1988
- 24 **Lance W. Small, Editor,** Noetherian rings and their applications, 1987
- 23 **E. H. Rothe,** Introduction to various aspects of degree theory in Banach spaces, 1986
- 22 **Michael E. Taylor,** Noncommutative harmonic analysis, 1986
- 21 **Albert Baernstein, David Drasin, Peter Duren, and Albert Marden, Editors,** The Bieberbach conjecture: Proceedings of the symposium on the occasion of the proof, 1986
- 20 **Kenneth R. Goodearl,** Partially ordered abelian groups with interpolation, 1986
- 19 **Gregory V. Chudnovsky,** Contributions to the theory of transcendental numbers, 1984
- 18 **Frank B. Knight,** Essentials of Brownian motion and diffusion, 1981
- 17 **Le Baron O. Ferguson,** Approximation by polynomials with integral coefficients, 1980
- 16 **O. Timothy O'Meara,** Symplectic groups, 1978
- 15 **J. Diestel and J. J. Uhl, Jr.,** Vector measures, 1977
- 14 **V. Guillemin and S. Sternberg,** Geometric asymptotics, 1977
- 13 **C. Pearcy, Editor,** Topics in operator theory, 1974
- 12 **J. R. Isbell,** Uniform spaces, 1964
- 11 **J. Cronin,** Fixed points and topological degree in nonlinear analysis, 1964
- 10 **R. Ayoub,** An introduction to the analytic theory of numbers, 1963
- 9 **Arthur Sard,** Linear approximation, 1963
- 8 **J. Lehner,** Discontinuous groups and automorphic functions, 1964
- 7.2 **A. H. Clifford and G. B. Preston,** The algebraic theory of semigroups, Volume II, 1961
- 7.1 **A. H. Clifford and G. B. Preston,** The algebraic theory of semigroups, Volume I, 1961
- 6 **C. C. Chevalley,** Introduction to the theory of algebraic functions of one variable, 1951
- 5 **S. Bergman,** The kernel function and conformal mapping, 1950
- 4 **O. F. G. Schilling,** The theory of valuations, 1950
- 3 **M. Marden,** Geometry of polynomials, 1949

(See the AMS catalog for earlier titles)

MATHEMATICAL
Surveys and Monographs

Volume 41

Monotone Dynamical Systems
An Introduction to the Theory of Competitive and Cooperative Systems

Hal L. Smith

American Mathematical Society
Providence, Rhode Island

The author was supported in part by NSF grant #DMS 9141550.

1991 *Mathematics Subject Classification.* Primary 34C11, 34C15, 34C25, 34C35, 34K15, 34K20, 35B40, 35B50, 35K55.

Library of Congress Cataloging-in-Publication Data
Smith, Hal L.
  Monotone dynamical systems: an introduction to the theory of competitive and cooperative systems / Hal L. Smith.
      p. cm. — (Mathematical surveys and monographs, ISSN 0076-5376; v. 41)
  Includes bibliographical references and index.
  ISBN 0-8218-0393-X
    1. Differentiable dynamical systems. 2. Monotonic functions. I. Title. II. Series: Mathematical surveys and monographs; no. 41.
QA614.8.S63 1995
515'.352–dc20   94-48032
                                                                                                  CIP

Copying and reprinting. Individual readers of this publication, and nonprofit libraries acting for them, are permitted to make fair use of the material, such as to copy a chapter for use in teaching or research. Permission is granted to quote brief passages from this publication in reviews, provided the customary acknowledgment of the source is given.
  Republication, systematic copying, or multiple reproduction of any material in this publication (including abstracts) is permitted only under license from the American Mathematical Society. Requests for such permission should be addressed to the Manager of Editorial Services, American Mathematical Society, P.O. Box 6248, Providence, Rhode Island 02940-6248. Requests can also be made by e-mail to reprint-permission@math.ams.org.
  The owner consents to copying beyond that permitted by Sections 107 or 108 of the U.S. Copyright Law, provided that a fee of $1.00 plus $.25 per page for each copy be paid directly to the Copyright Clearance Center, Inc., 222 Rosewood Drive, Danvers, Massachusetts 01923. When paying this fee please use the code 0076-5376/95 to refer to this publication. This consent does not extend to other kinds of copying, such as copying for general distribution, for advertising or promotional purposes, for creating new collective works, or for resale.

© Copyright 1995 by the American Mathematical Society. All rights reserved.
Printed in the United States of America.
The American Mathematical Society retains all rights
except those granted to the United States Government.
⊚ The paper used in this book is acid-free and falls within the guidelines
established to ensure permanence and durability.
♻ Printed on recycled paper.
This volume was printed directly from copy prepared by the author using $\mathcal{A}_{\mathcal{M}}\mathcal{S}$-TEX,
the American Mathematical Society's TEX macro system.

Contents

Preface	vii
Chapter 1. Monotone Dynamical Systems	1
0. Introduction	1
1. Definitions and Preliminary Results	1
2. The Convergence Criterion	3
3. The Limit Set Dichotomy	5
4. Quasiconvergence is Generic	8
5. Remarks and Discussion	13
Chapter 2. Stability and Convergence	15
0. Introduction	15
1. Stability	15
2. The Order Interval Trichotomy	16
3. Some Global Results	18
4. Generic Convergence to Equilibrium	19
5. Unstable Equilibria and Connecting Orbits	24
6. Remarks and Discussion	28
Chapter 3. Competitive and Cooperative Differential Equations	31
0. Introduction	31
1. The Kamke Condition	32
2. Positively Invariant Sets and Monotone Solutions	34
3. Main Results	37
4. Three Dimensional Systems	40
5. Alternative Cones	48
6. The Field-Noyes Model	50
7. Remarks and Discussion	51
Chapter 4. Irreducible Cooperative Systems	55
0. Introduction	55
1. Strong Monotonicity	56
2. A Biochemical Control Circuit	58
3. Stability and the Perron-Frobenius Theorem	60
4. Competition and Migration	64
5. Smale's Construction	71
6. Remarks and Discussion	72

Chapter 5. Cooperative Systems of Delay Differential Equations ... 75
   0. Introduction ... 75
   1. The Quasimonotone condition ... 78
   2. Positively Invariant Sets, Monotone Solutions, and Contracting Rectangles ... 81
   3. Eventual Strong Monotonicity ... 85
   4. Generic Convergence for Cooperative and Irreducible Systems ... 89
   5. Stability of Equilibria ... 91
   6. A Biochemical Control Circuit with Delays ... 93
   7. Competition with Time Delays ... 94
   8. Remarks and Discussion ... 98

Chapter 6. Nonquasimonotone Delay Differential Equations ... 101
   0. Introduction ... 101
   1. The Exponential Ordering ... 101
   2. The Strong Order Preserving Property ... 107
   3. Generic Convergence to Equilibrium ... 109
   4. Stability of Equilibria ... 110
   5. A Model of an Adult Fly Population ... 112
   6. Remarks and Discussion ... 116

Chapter 7. Quasimonotone Systems of Parabolic Equations ... 119
   0. Introduction ... 119
   1. Parabolic Systems: The Basic Setup ... 120
   2. Maximum Principles ... 123
   3. Positively Invariant Sets, Comparison and Monotonicity ... 127
   4. The Strong Order Preserving Property ... 132
   5. Generic Convergence for Cooperative and Irreducible Systems ... 134
   6. Stability of Equilibria ... 136
   7. The Biochemical Control Circuit with Diffusion ... 139
   8. Remarks and Discussion ... 141

Chapter 8. A Competition Model ... 145
   0. Introduction ... 145
   1. The Model ... 145
   2. A Single Population ... 151
   3. Stability of the Equilibria E_0, E_1, E_2 ... 154
   4. Coexistence ... 158
   5. Remarks and Discussion ... 161

Appendix. Chain Recurrence ... 163

Bibliography ... 167

Index ... 173

Preface

There is a long history of the application of monotone methods and comparison arguments in differential equations. The early works of Müller(1926) and Kamke(1932) laid the foundation at least for ordinary differential equations. The work of Krasnoselskii(1964,1968) has been especially influential. However, not until the work of M.W.Hirsch were monotonicity methods fully integrated with dynamical systems ideas. In a remarkable series of papers entitled "Systems of differential equations that are competitive or cooperative", parts I through VI, Hirsch develops what is now often referred to as monotone dynamical systems theory. There, he shows that the generic solution of a cooperative and irreducible system of ordinary differential equations converges to the set of equilibria. Furthermore, the flow on a compact limit set of an n-dimensional cooperative or competitive system of differential equations is shown to be topologically conjugate to the flow of an $n-1$-dimensional system of differential equations, restricted to a compact invariant set. The Poincaré-Bendixson Theorem is established for three dimensional competitive or cooperative systems. The theory of infinite dimensional dynamical systems that preserve a partial order relation is developed in the important work Hirsch(1988b). Here one finds for instance that the generic orbit of such a system is stable and converges to the set of equilibria. Applications are made to reaction-diffusion systems.

The infinite dimensional theory of monotone systems has been heavily influenced by the work of H.Matano. The beginnings of the theory appear in Matano (1979) which focuses on semilinear diffusion equations. Matano(1984) introduces the important idea of a strongly order preserving semiflow. This condition is more flexible than strong monotonicity, used by Hirsch, because it does not require the order cone to have nonempty interior. The main result of this paper establishes the existence of monotone heteroclinic orbits and has motivated considerable research in this direction. Matano(1986) and (1987) contain outlines, without proof, of extensions of results of Hirsch using the strong order preserving property in place of strong monotonicity.

The work of Smith and Thieme(1990,1991) represents a synthesis of the approaches of Hirsch and Matano that attempts to simplify and streamline the arguments. Significant improvements in the theory are obtained with additional compactness hypotheses that are usually satisfied in the applications.

There presently exists no current account of the important results in the theory of continuous-time monotone dynamical systems and particularly, of competitive and cooperative systems of differential equations. The earlier survey Smith(1988) is now somewhat dated and, in any case, treats primarily cooperative systems. In view of the many existing and potential applications of this theory, there seems to be a need for some overview of the topic. The present monograph is intended to serve that purpose. However, it is not intended as a comprehensive survey of the literature on monotone dynamical systems. In particular, the references do not include all the important work in this area. Nor has any attempt been made to present the most general results possible. In fact, many of the results presented here are special cases of more general ones that appear in the literature. Generality is readily sacrificed for simplicity and clarity of presentation. As the title suggests, this work is an introduction to the subject of monotone dynamics and competitive and cooperative systems. The selection of material presented here reflects a personal bias in favor of those results which are useful in applications. Some results appear here for the first time and new and simpler proofs have been found in a number of cases.

This monograph is an expanded version of a series of talks given by the author at National Tsing Hua University, Taiwan, at the invitation of Professor S.B.Hsu, during August, 1993, and supported by the National Council of Science, Republic of China. The author would like to especially thank Professor Hsu for the invitation to visit Taiwan and for his careful attention to these talks.

Many of the results presented in this monograph represent joint work with Horst R. Thieme. The author has benefited greatly from this collaboration over the years and this work would not have been possible without it. The author would also like to acknowledge the influence of his teacher and collaborator, Paul Waltman, on this work. Some of the most beautiful applications of the theory of monotone systems appear in the papers of Waltman and various collaborators. Courage to take on this project was derived from the recent collaboration of the author with Paul Waltman on a monograph titled "The Theory of the Chemostat". Finally, the author's collaborations with Robert H. Martin have greatly influenced the presentation of Chapter 7 on application of the theory to parabolic partial differential equations.

This monograph does not treat discrete dynamical systems generated by monotone mappings, an area which has flourished in recent years. Although Hirsch's work includes some results in this area, a different set of players are responsible for the major results. The book of Hess(1991) is an excellent reference for this work, as are the papers of N.Dancer and P.Hess and of P.Takac. As the reader might guess this theory is more difficult (there is no Limit Set Dichotomy) and, it seems, monotonicity does not restrict the dynamics of maps as severely as it does for flows.

In order to present the results in a clear way, we have tried to avoid references to the literature in the main body of each chapter. A separate section titled "Remarks and Discussion" is aimed at providing the reader with references to the literature, citations for the main results, a discussion of related results, and extensions and applications of the theory. This is where the reader will find references to more general results. A brief description of the contents of each chapter follows.

Chapter 1 develops the main tools of monotone dynamical systems theory in ordered metric spaces. These include the Convergence Criteria, the Limit Set Dichotomy and the Sequential Limit Set Trichotomy. The main result of the chapter is that, under suitable hypotheses, a generic orbit of a monotone semiflow converges to the set of equilibria.

Chapter 2 begins with results on stability and the classical order interval trichotomy. Results on global convergence and sufficient conditions for all orbits to approach the set of equilibria are formulated. The main result of this chapter establishes sufficient conditions for the generic orbit of a monotone semiflow to converge to a single equilibrium. Here, the Krein-Rutman theorem is a principal tool. Sub-equilibrium and super-equilibrium are introduced in the final section and sufficient conditions are given for the existence of a monotone heteroclinic orbit.

Chapter 3 is essentially independent of the first two chapters, requiring only the convergence criterion (theorem 1.2) from Chapter 1. The reader who is primarily interested in ordinary differential equations can begin here. The focus is on competitive and cooperative systems of ordinary differential equations which are not necessarily irreducible. The Kamke condition is shown to imply monotonicity of the forward flow. The centerpiece of this chapter is the result that the flow on a compact limit set of an n-dimensional competitive or cooperative system is topologically equivalent to the flow of a Lipschitz $n-1$-dimensional system restricted to a compact invariant set. This then leads to the result that all solutions of planar competitive and cooperative systems converge to equilibrium and to the Poincaré-Bendixson theorem for three dimensional systems. An application is made to the Field-Noyes model of the Belousov Zhabotinskii reaction.

In Chapter 4, the irreducibility hypothesis is exploited for competitive and cooperative systems. It is shown to imply that the forward flow of a cooperative and irreducible system is strongly monotone. The generic convergence result of Chapter 2 is translated into the language of cooperative irreducible systems of ordinary differential equations. It is applied to two examples, a protein synthesis model in physiology and a model of two populations competing in an environment consisting of two habitats between which each population can migrate. The Perron-Frobenius theorem and its implications for the stability of steady states and periodic solutions is discussed. The important construction of Smale showing that any dynamics can be realized within the class of competitive and cooperative systems of sufficiently large dimension is described in detail.

The results of Chapter 1 and 2 are applied to quasimonotone delay differential equations in Chapter 5. The class of cooperative and irreducible delay systems is identified and the generic orbit of such a system is shown to converge to equilibrium. To each such system one can associate a cooperative and irreducible system of ordinary differential equations which has essentially the same dynamics as the system of functional differential equations. An application is given to the protein synthesis model with delays. We also show in this chapter that if a delay differential equation, which is not necessarily quasimonotone, possesses a positively invariant rectangle then one can associate to it two comparison systems of quasimonotone delay differential equations whose solutions can be compared to the original system. Therefore, quasimonotone systems arise naturally from systems of delay equations which are not quasimonotone but have a positively invariant rectangle.

In Chapter 6 the so-called exponential ordering on the space of continuous functions is introduced and shown to allow a weakening of the rather restrictive quasimonotone condition of chapter 5 for delay differential equations. The treatment follows Smith and Thieme(1991). A surprising outcome of the results of this chapter is that delay differential equations with delays which are small compared to the Lipschitz constant always generate monotone semiflows in the exponential ordering. As a result, convergence to equilibrium is generic. Application is made to a model of a fly population.

The focus of Chapter 7 is on systems of reaction diffusion equations satisfying the quasimonotone condition. It is shown that such systems define a semiflow on a subspace of the space of continuous functions on a bounded domain. A section is devoted to maximum principles, the main tool in the analysis of these systems. Monotonicity, the existence of positively invariant sets and comparison results are developed. The main result provides sufficient conditions for the generic solution to converge to equilibrium. The stability of equilibria is examined using the Krein-Rutman Theorem and an application is given to the biochemical control circuit.

The results of Chapter 7 are applied to a model of microbial competition in the chemostat with convection and diffusion in the final Chapter 8.

Some remarks on format are necessary. As the table of contents suggests, each chapter is divided into numbered sections. Results, whether they be theorems, propositions, or lemmas, are numbered consecutively within each section. For example, the third result of section 2 of Chapter 5 is labeled theorem 2.3 (or proposition 2.3 or lemma 2.3) within the text of Chapter 5. When referring to that same result in another chapter, we write theorem 5.2.3. New terms are italicized the first time they appear in the text.

The reader who is primarily interested in the applications to ordinary differential equations may read the first two sections of Chapter 1 for the basic definitions and notation and skim the main results of Chapters 1 (theorem 1.4.3) and Chapter 2 (theorem 2.4.7) before proceeding to Chapters 3 and 4.

Several colleagues read various portions of the manuscript for this monograph and provided valuable suggestions and comments. The author wishes to express his sincere thanks for their effort: C. Cosner, R.H. Martin, J.Wu, W.E. Fitzgibbon.

Finally, the author wishes to thank Linda Arneson for producing a beautiful type-set manuscript to AMS standards and Bruce Long for rendering most of the figures in the text.

CHAPTER 1

Monotone Dynamical Systems

0. Introduction

The aim of this chapter is to introduce the basic definitions and develop the main tools of the theory of monotone dynamical systems. Of course, the term "monotone" suggests the existence of some sort of order relation so we begin by describing our basic setting, a partially ordered metric space. This is the space of "states" and the dynamical system on this space is described by a semiflow, that is by a mapping taking an initial state at time zero to the state of the system at a subsequent time. The standard dynamical systems ideas are briefly recalled. A monotone dynamical system is just a dynamical system on an ordered metric space which has the property that ordered initial states lead to ordered subsequent states. Actually, a stronger property than monotonicity, the so-called strong order preserving property, is required for many of the results presented here. Following these basic definitions, the development of the theory begins. The main result of this chapter, Theorem 4.3, provides sufficient conditions for the omega limit set of a generic point of a strongly order preserving dynamical system to be contained in the set of equilibria. However, important results are obtained along the way such as the Convergence Criterion, the Nonordering of Limit Sets, the Limit Set Dichotomy and the Sequential Limit Set Trichotomy. Dynamical systems that possess only the monotonicity property cannot have attracting periodic orbits, as shown by Theorem 2.2.

Section 3, leading to the Limit Set Dichotomy, is the most technical part of this chapter and might be skimmed over lightly on a first reading. Ultimately, however, it is this result more than any other that is the key to the theory.

1. Definitions and Preliminary Results

Let X be an ordered metric space with metric d and *partial order* relation $\leq$. Recall that a partial order relation satisfies:
(i) reflexive: $x \leq x$ for all $x \in X$;
(ii) transitive: $x \leq y$ and $y \leq z$ implies $x \leq z$;
(iii) antisymmetric: $x \leq y$ and $y \leq x$ implies $x = y$.

We write $x < y$ if $x \leq y$ and $x \neq y$. Given two subsets A and B of X, we write $A \leq B$ ($A < B$) when $x \leq y$ ($x < y$) holds for each choice of $x \in A$ and $y \in B$.

We assume that the order relation and the topology on X are compatible in the sense that $x \leq y$ whenever $x_n \to x$ and $y_n \to y$ as $n \to \infty$ and $x_n \leq y_n$ for all n. This is just to say that the partial order relation is closed. For $A \subset X$ we write $\bar{A}$ for the closure of A and Int$A$ for the interior of A.

In the applications, X is typically a subset of a Banach space Y with a *positive cone* Y_+. That is, Y_+ is a nonempty closed subset of Y with the properties:

1

(1) $\mathbb{R}_+ \cdot Y_+ \subset Y_+$,
(2) $Y_+ + Y_+ \subset Y_+$,
(3) $Y_+ \cap (-Y_+) = 0$.

In this case, the relation defined by $x \leq y$ if and only if $y - x \in Y_+$ is a closed partial order relation.

The most important examples for our purposes are Banach spaces of real-valued functions on some set Ω where Y_+ is the subset of those functions which take only non-negative values for all (or almost all) points of Ω. For example, when $\Omega = \{1, 2, \ldots, n\}$, then the space Y becomes $\mathbb{R}^n$ with cone $Y_+ = \mathbb{R}^n_+$, the set of vectors with non-negative components. The corresponding ordering $x \leq y$ means that $x_i \leq y_i$ holds for all i. If Y is the Banach space of continuous functions on a compact topological space Ω and Y_+ is the cone of non-negative functions, then the ordering $x \leq y$ means that $x(t) \leq y(t)$ for all $t \in \Omega$.

A *semiflow* on X is a continuous map $\Phi : X \times \mathbb{R}^+ \to X$ which satisfies:
(i) $\Phi_0 = id_X$.
(ii) $\Phi_t \circ \Phi_s = \Phi_{t+s}$ for $t, s \geq 0$.

Here, $\Phi_t(x) \equiv \Phi(x, t)$ for $x \in X$ and $t \geq 0$ and id_X is the identity map on X. The *orbit* of x is denoted by $O(x)$

$$O(x) = \{\Phi_t(x) : t \geq 0\}.$$

An *equilibrium point* is a point x for which $O(x) = \{x\}$. Let E be the set of all equilibrium points for Φ. A subset A of X is *positively invariant* if $\Phi_t A \subset A$ for all $t \geq 0$. It is *invariant* if $\Phi_t A = A$ for all $t \geq 0$. $O(x)$ is said to be a *T-periodic orbit* for some $T > 0$ if $\Phi_T(x) = x$. In that case, $\Phi_{t+T}(x) = \Phi_t(x)$ for all $t \geq 0$ so $O(x) = \{\Phi_t(x) : 0 \leq t \leq T\}$. We call $\Phi_t(x)$ a *T-periodic solution*. A T-periodic solution can be extended to $\mathbb{R}$ as the unique T-periodic function $u(t)$ which agrees with $\Phi_t(x)$ for $0 \leq t \leq T$. It is a solution in the sense that $\Phi_t(u(s)) = u(t + s)$ for $t \geq 0$ and $s \in \mathbb{R}$.

The *omega limit set*, $\omega(x)$, of $x \in X$ is defined by

$$\omega(x) = \cap_{t \geq 0} \overline{\cup_{s \geq t} \Phi_s(x)}.$$

If $\overline{O(x)}$ is compact then $\omega(x)$ is nonempty, compact, connected, invariant and attracts $\Phi_t(x)$, that is, dist$(\omega(x), \Phi_t(x)) \to 0$ as $t \to \infty$ (see e.g. Saperstone(1981)). Recall that dist$(D, x) = \inf_{y \in D} d(y, x)$ gives the distance of a point $x \in X$ from a subset D of X.

A point $x \in X$ is called a *quasiconvergent point* if $\omega(x) \subset E$. The set of all such points is denoted by Q. A point x is called a *convergent point* if $\omega(x)$ consists of a single point of E. The set of all convergent points is denoted by C.

The semiflow Φ is said to be *monotone* provided

$$\Phi_t(x) \leq \Phi_t(y) \text{ whenever } x \leq y \text{ and } t \geq 0.$$

Φ is called *strongly order-preserving*, SOP for short, if it is monotone and whenever $x < y$ there exist open subsets U, V of X with $x \in U$ and $y \in V$ and $t_0 > 0$ such that

$$\Phi_{t_0} U \leq \Phi_{t_0} V.$$

Monotonicity of Φ then implies that $\Phi_t U \leq \Phi_t V$ for all $t \geq t_0$.

If the partial order relation is generated by a cone Y_+ in a Banach space Y and if Y_+ has nonempty interior, IntY_+, in Y then we write $x \ll y$ whenever

$y - x \in \text{Int} Y_+$. This is a stronger relation than "$\leq$", that is $x \ll y$ implies $x < y$. As addition in Y is a continuous map from $Y \times Y \to Y$, it follows that if $x \ll y$ then there exists open neighborhoods U of x and V of y such that $V - U \subset \text{Int} Y_+$. Consequently, $U < V$.

Suppose that the partial order on X is generated by the cone Y_+ which has nonempty interior, $\text{Int} Y_+$. The semiflow Φ is said to be *strongly monotone* on X ($X \subset Y$) if Φ is monotone and whenever $x < y$ and $t > 0$ then $\Phi_t(x) \ll \Phi_t(y)$. Φ is called *eventually strongly monotone* if it is monotone and whenever $x < y$ there exists $t_0 > 0$ such that $\Phi_{t_0}(x) \ll \Phi_{t_0}(y)$. Clearly, if Φ is strongly monotone then it is eventually strongly monotone.

PROPOSITION 1.1. *If Φ is eventually strongly monotone, then it is SOP.*

PROOF. If $x < y$, then there exists $t_0 > 0$ such that $\Phi_{t_0}(x) \ll \Phi_{t_0}(y)$. Take neighborhoods $\tilde{U}$ of $\Phi_{t_0}(x)$ and $\tilde{V}$ of $\Phi_{t_0}(y)$ such that $\tilde{U} < \tilde{V}$. This can be done since $\Phi_{t_0}(x) \ll \Phi_{t_0}(y)$. By continuity of Φ_{t_0}, there are neighborhoods U of x and V of y such that $\Phi_{t_0}(U) \subset \tilde{U}$ and $\Phi_{t_0}(V) \subset \tilde{V}$. Therefore, $\Phi_{t_0}(U) < \Phi_{t_0}(V)$ as required. □

REMARK 1.1. In many applications of the theory, the SOP condition can be verified by showing that Φ is eventually strongly monotone. However, it sometimes occurs that Y_+ has empty interior in Y. In this case, the strategy is to find a Banach space Z such that $Z \subset Y$, the inclusion map $i: Z \to Y$ is a continuous linear operator, and $Z_+ \equiv Z \cap Y_+$ has nonempty interior, $\text{Int} Z_+$, in Z. Observe that Z_+ is a cone in Z. If it can be shown that whenever $x < y$ holds in X there is a $t_0 > 0$ such that $\Phi_{t_0}(x) \ll \Phi_{t_0}(y)$ holds in Z and if Φ_{t_0} maps X continuously into Z then the arguments of the previous proof easily extend to show that Φ is SOP.

The following well-known result (e.g. Hess(1991) Lemma 1.1) is basic.

LEMMA 1.2. *A monotone sequence contained in a compact subset of X converges in X.*

PROOF. Suppose that x_n is a sequence satisfying $x_n \leq x_{n+1}$ and $x_n \in A$ for $n \geq 1$ where A is a compact subset of X; the case of a decreasing sequence is treated similarly. It follows that the sequence x_n has convergent subsequences. The Lemma will be proved by showing that there exists a unique $p \in X$ which is the limit of every convergent subsequence. If x_{n_k} and x_{m_k} are two subsequences of x_n and if $x_{n_k} \to p$ and $x_{m_k} \to q$ as $k \to \infty$ then, by monotonicity of x_n, for each k there exists $l(k)$ such that $x_{n_k} \leq x_{m_{l(k)}}$. Passing to the limit as $k \to \infty$, $p \leq q$. A similar argument shows that $q \leq p$. By the antisymmetry property (iii) of the order relation, $q = p$. □

2. The Convergence Criterion

Several fundamental results in the theory of monotone dynamical systems are proved in this section. All are based on the following sufficient conditions for a solution to converge to equilibrium. Hereafter, we assume that Φ is monotone and $\overline{O(x)}$ is a compact subset of X for each $x \in X$.

THEOREM 2.1. *Convergence Criterion. Let $\Phi_T(x) \geq x$ for some $T > 0$. Then $\omega(x)$ is a T-periodic orbit. If $\Phi_t(x) \geq x$ for t belonging to some nonempty open* subset of $(0, \infty)$ then $\Phi_t(x) \to p \in E$ as $t \to \infty$. In particular, if Φ is SOP and $\Phi_T(x) > x$ for some $T > 0$ then $\Phi_t(x) \to p \in E$ as $t \to \infty$.

PROOF. Monotonicity implies that $\Phi_{(n+1)T}(x) \geq \Phi_{nT}(x)$ for $n = 1, 2, \ldots$ and therefore $\Phi_{nT}(x) \to p$ as $n \to \infty$ for some p by the compactness of the orbit closure. By continuity of Φ,

$$\Phi_{t+T}(p) = \Phi_{t+T}(\lim_{n \to \infty} \Phi_{nT}(x))$$
$$= \lim_{n \to \infty} \Phi_{(n+1)T+t}(x)$$
$$= \lim_{n \to \infty} \Phi_t(\Phi_{(n+1)T}(x))$$
$$= \Phi_t(p)$$

for all $t \geq 0$. Hence, $O(p)$ is a $T-$ periodic orbit. If $t_j \to \infty$ and $\Phi_{t_j}(x) \to q$, $j \to \infty$, write $t_j = n_j T + r_j$ where n_j is a natural number and $0 \leq r_j < T$. By passing to a subsequence if necessary, we may assume that $r_j \to r$ as $j \to \infty$. Since $n_j \to \infty$ as $j \to \infty$, we have

$$\Phi_{t_j}(x) = \Phi_{r_j}(\Phi_{n_j T}(x)) \to \Phi_r(p) = q$$

as $j \to \infty$ where $0 \leq r \leq T$. Therefore, $\omega(x) = O(p)$. This proves the first assertion.

If $\Phi_t(x) \geq x$ holds for $t \in (T - \epsilon, T + \epsilon)$ for some $T > 0$ and $0 < \epsilon < T$ then $\omega(x) = O(p) = \{\Phi_t(p) : 0 \leq t \leq T\}$ where $\Phi_{nT}(x) \to p$ as $n \to \infty$ and $O(p)$ is a $T-$periodic orbit, by the previous argument. Applying the same argument with $\tau \in (T - \epsilon, T + \epsilon)$ replacing T we find that $\omega(x)$ is a $\tau-$periodic orbit. But $\omega(x) = O(p)$ so it follows that

$$\Phi_{t+\tau}(p) = \Phi_t(p)$$

for all $t \geq 0$ (the reader should verify this). Therefore, $\Phi_t(p)$ is $\tau-$periodic for every $\tau \in (T - \epsilon, T + \epsilon)$. Let G be the set of all periods of $\Phi_t(p)$. G is closed under addition and contains the interval $(T - \epsilon, T + \epsilon)$. If $0 \leq s < \epsilon$ and $t \geq 0$ then

$$\Phi_{t+s}(p) = \Phi_t(\Phi_s(p)) = \Phi_t(\Phi_{s+T}(p)) = \Phi_t(p).$$

Hence, $[0, \epsilon) \subset G$ and therefore $G = \mathbb{R}^+$ and $p \in E$. This proves the second assertion.

If $\Phi_T(x) > x$ and Φ is SOP then there exist neighborhoods U of x and V of $\Phi_T(x)$ and $t_0 > 0$ such that $\Phi_{t_0}(U) \leq \Phi_{t_0}(V)$. It follows that $\Phi_{t_0}(x) \leq \Phi_{t_0+T+\epsilon}(x)$ for all ϵ sufficiently small. The previous assertion implies the $\omega(x) = p \in E$. □

Among the many consequences of the Convergence Criterion is the result that a monotone dynamical system cannot have an *attracting periodic orbit*. A periodic orbit O is attracting if there is an open set U containing O with the property that $\omega(x) = O$ for every $x \in U$. In order to prove this result, we must require that the space X have the following property:

(P) Given $p \in X$ and a neighborhood U of p in X, there exists a neighborhood V of p, $V \subset U$, and $y \in U$ such that either $y \leq V$ or $V \leq y$.

Hypothesis (P) holds if $X \subset Y$, the partial order on X is generated by a cone Y_+ with nonempty interior in the Banach space Y and if every neighborhood U of

p contains a point y satisfying either $y \ll p$ or $p \ll y$. For example, if X is open in Y this will hold.

THEOREM 2.2. *If (P) holds then there does not exist a non-trivial, attracting periodic orbit.*

PROOF. Let $O = \{\Phi_t(p) : 0 \leq t \leq T\}$ be an attracting T–periodic orbit. Then there exists a neighborhood U of p such that if $x \in U$ then $\omega(x) = O$. By (P), there exists a neighborhood V of p contained in U and $y \in U$ such that $y \leq V$ or $V \leq y$. We assume the former holds as the argument is similar in both cases. There exists $t_j \to \infty$ such that $\Phi_{t_j}(y) \to p$ so $\Phi_{t_j}(y)$ belongs to V for all large j. Fix such a j, then $y \leq \Phi_{t_j}(y)$. Since V is open and Φ is continuous, $\Phi_{t_j+s}(y) \in V$ for all small s so $y \leq \Phi_{t_j+s}(y)$ for small s. By the Convergence Criterion, $\omega(y) = O$ is a single element of E. This is a contradiction to our assumption that O is a nontrivial periodic orbit. $\square$

The next result describes how an omega limit set is imbedded in the space X. It is fundamental to the theory.

THEOREM 2.3. *Nonordering of Limit Sets. An omega limit set cannot contain distinct points x and y with the property that there exists neighborhoods U of x and V of y such that $U \leq V$. If Φ is SOP then a limit set cannot contain two points x and y with $x < y$.*

PROOF. Suppose that $\omega(z)$ contains distinct points x and y possessing neighborhoods U and V, respectively, such that $U \leq V$. $\Phi_{t_1}(z) \in U$ for some $t_1 > 0$. Choose t_2 greater than t_1 such that $\Phi_{t_2}(z) \in V$. It follows that $\Phi_t(z) \in V$ for all t sufficiently near to t_2 and for these t we have

$$\Phi_t(z) = \Phi_{t-t_1}\,\Phi_{t_1}(z) \geq \Phi_{t_1}(z).$$

By the second assertion of the Convergence Criterion, $\Phi_t(z) \to p \in E$ as $t \to \infty$ so $\omega(z) = p$, a contradiction.

If $x < y$, $x, y \in \omega(z)$ and Φ is SOP, then there exists neighborhoods U of x and V of y and $t_0 > 0$ such that $\Phi_{t_0}(U) \leq \Phi_{t_0}(V)$. Choose $t_1 > 0$ such that $\Phi_{t_1}(z) \in U$ and $t_2 > t_1$ such that $\Phi_{t_2}(z) \in V$. As $\Phi_t(z) \in V$ for all t near t_2, it follows that $\Phi_{t_1+t_0}(z) \leq \Phi_{t+t_0}(z) = \Phi_{t-t_1}\,\Phi_{t_1+t_0}(z)$ holds for all t near t_2. By the Convergence Criterion, $\Phi_t(z) \to p \in E$ as $t \to \infty$ so $\omega(z) = p$, a contradiction. $\square$

An immediate consequence of the Nonordering of Limit Sets is that an omega limit set cannot contain a maximal (minimal) element.

COROLLARY 2.4. *If Φ is SOP, $a \in \omega(x)$ and $\omega(x) \leq a$ ($a \leq \omega(x)$), then $\omega(x) = \{a\}$.*

PROOF. Suppose $a \in \omega(x)$ and $\omega(x) \leq a$. If $b \in \omega(x)$ and $b \neq a$ then $b < a$. This violates the last assertion of the previous result. $\square$

3. The Limit Set Dichotomy

Throughout the remainder of this chapter, we assume that Φ is SOP. Our immediate goal in this section is to prove the Limit Set Dichotomy: If $x < y$ then either (a) $\omega(x) < \omega(y)$, or (b) $\omega(x) = \omega(y) \subset E$. A number of preliminary results are required.

PROPOSITION 3.1. *Colimiting Principle.* If $x < y$, $t_k \to \infty$, $\Phi_{t_k}(x) \to p$ and $\Phi_{t_k}(y) \to p$ as $k \to \infty$ then $p \in E$.

PROOF. Choose neighborhoods U of x and V of y and $t_0 > 0$ such that $\Phi_{t_0}(U) \leq \Phi_{t_0}(V)$. Let $\delta > 0$ be so small that $\{\Phi_s(x) : 0 \leq s \leq \delta\} \subset U$ and $\{\Phi_s(y) : 0 \leq s \leq \delta\} \subset V$. Then $\Phi_s(x) \leq \Phi_r(y)$ whenever $t_0 \leq r, s \leq t_0 + \delta$. Therefore,
$$\Phi_{t_k - t_0}(\Phi_s(x)) \leq \Phi_{t_k - t_0}(\Phi_{t_0}(y)) = \Phi_{t_k}(y) \tag{3.1}$$
for all $s \in [t_0, t_0 + \delta]$ and all large k. As
$$\Phi_{t_k - t_0}(\Phi_s(x)) = \Phi_{s - t_0}(\Phi_{t_k}(x)) = \Phi_r(\Phi_{t_k}(x))$$
where $r = s - t_0 \in [0, \delta]$ if $s \in [t_0, t_0 + \delta]$, we have
$$\Phi_r(\Phi_{t_k}(x)) \leq \Phi_{t_k}(y)$$
for large k and $r \in [0, \delta]$. Passing to the limit as $k \to \infty$ we find that $\Phi_r(p) \leq p$ for $0 \leq r \leq \delta$. If, in (3.1) above, we replace $\Phi_s(x)$ by $\Phi_{t_0}(x)$ and replace $\Phi_{t_0}(y)$ by $\Phi_s(y)$, and argue as above then we find that $p \leq \Phi_r(p)$ for $0 \leq r \leq \delta$. Evidently, $\Phi_r(p) = p$, $0 \leq r \leq \delta$ and therefore for all $r \geq 0$, so $p \in E$. □

PROPOSITION 3.2. *Intersection Principle.* If $x < y$ then $\omega(x) \cap \omega(y) \subset E$.

PROOF. If $p \in \omega(x) \cap \omega(y)$ then there exists a sequence $t_k \to \infty$ such that $\Phi_{t_k}(x) \to p$. By passing to a subsequence if necessary, we can assume that $\Phi_{t_k}(y) \to q$. Monotonicity implies that $p \leq q$. If $p < q$ then we contradict the Nonordering of Limit Sets since $p, q \in \omega(y)$. Hence $p = q$. The Colimiting Principle then implies that $p \in E$. □

REMARK 3.1. The proof shows that if $x < y$ and $p \in \omega(x) \cap \omega(y)$, then $\Phi_{t_k}(x) \to p$ if and only if $\Phi_{t_k}(y) \to p$ for a sequence $t_k \to \infty$.

LEMMA 3.3. *Let K_1 and K_2 be two compact subsets of X satisfying $K_1 < K_2$. Then there are open sets U and V, $K_1 \subset U$, $K_2 \subset V$, and $t_1 \geq 0, \epsilon > 0$ such that*
$$\Phi_{t+s}(U) \leq \Phi_t(V), \quad t \geq t_1, \ 0 \leq s \leq \epsilon.$$

PROOF. Let $x \in K_1$. For each $y \in K_2$ there exists $t_y \geq 0$ and a neighborhood U_y of x and a neighborhood V_y of y such that $\Phi_t(U_y) \leq \Phi_t(V_y)$ for $t \geq t_y$ since Φ is strongly order preserving. Then $\{V_y\}_{y \in K_2}$ is an open cover of K_2, so we may choose a finite subcover: $K_2 \subset \cup_{i=1}^n V_{y_i} \equiv \tilde{V}$ where $y_i \in K_2$, $1 \leq i \leq n$. Let $\tilde{U} = \cap_{i=1}^n U_{y_i}$, so $\tilde{U}$ is a neighborhood of x and let $\tilde{t} = \max_{1 \leq i \leq n} t_{y_i}$. Then, for each i, $\Phi_{\tilde{t}}(\tilde{U}) \subset \Phi_{\tilde{t}}(U_{y_i}) \leq \Phi_{\tilde{t}}(V_{y_i})$, so $\Phi_t(\tilde{U}) \leq \Phi_t(V_{y_i})$ for each i, $1 \leq i \leq n$, and $t \geq \tilde{t}$. It follows that $\Phi_t(\tilde{U}) \leq \Phi_t(\tilde{V})$ for $t \geq \tilde{t}$. For the remainder of the proof we write $\tilde{U}_x = \tilde{U}$ and $\tilde{V}_x = \tilde{V}$ to emphasize the dependence of these open sets on the point $x \in K_1$. Similarly $\tilde{t} = \tilde{t}_x$. Since $x \in K_1$ was arbitrary, we find for each $x \in K_1$ an open neighborhood $\tilde{U}_x$ of x, an open neighborhood $\tilde{V}_x$ of K_2 and $\tilde{t}_x \geq 0$ such that $\Phi_t(\tilde{U}_x) \leq \Phi_t(\tilde{V}_x)$ for $t \geq \tilde{t}_x$. Again, $\{\tilde{U}_x\}_{x \in K_1}$ is an open cover of K_1, so we extract a finite subcover $\tilde{U}_{x_i}$, $1 \leq i \leq m$, and set $U = \cup_{i=1}^m \tilde{U}_{x_i} \supset K_1$, $V = \cap_{i=1}^m \tilde{V}_{x_i} \supset K_2$ and $t_1 = \max_{1 \leq i \leq m} \tilde{t}_{x_i}$. Since $V \subset \tilde{V}_{x_i}$, $\Phi_t(\tilde{U}_{x_i}) \leq \Phi_t(V)$ for $t \geq t_1$, for each i, so $\Phi_t(U) \leq \Phi_t(V)$ for $t \geq t_1$.

In order to obtain the stronger conclusion of the lemma, note that for each $x \in K_1$ there exists $\epsilon_x > 0$ and a neighborhood W_x of x such that $\Phi([0, \epsilon_x) \times W_x) \subset U$. As $\{W_x\}_{x \in K_1}$ is an open cover of K_1, there exists $x_1, x_2, \ldots, x_m \in K_1$ such that $K_1 \subset \cup_{i=1}^m W_{x_i}$. Let $U' = \cup_{i=1}^m W_{x_i}$ and $\epsilon = \min_{1 \leq i \leq m} \epsilon_{x_i}$. If $x \in U'$ and $0 \leq s < \epsilon$ then $x \in W_{x_i}$ for some i so $\Phi_s(x) \in U$. Thus $\Phi([0, \epsilon) \times U') \subset U$ so $\Phi_s(U') \subset U$, $0 \leq s < \epsilon$. It follows that $\Phi_{t+s}(U') \subset \Phi_t(U) \leq \Phi_t(V)$ for $t \geq t_1$, $0 \leq s < \epsilon$. □

PROPOSITION 3.4. *Let x, y satisfy $x < y$. If $t_k \to \infty$, $\Phi_{t_k}(x) \to a$, $\Phi_{t_k}(y) \to b$ as $k \to \infty$ and $a < b$ then $O(a) < b$.*

PROOF. For $u \in \overline{O(x)}$, $v \in \overline{O(y)}$, $u \leq v$ define
$$\mathcal{J}(u, v) = \sup\{r \geq 0 : \Phi_t(u) \leq v, 0 \leq t \leq r\}.$$
The aim is to show that $\mathcal{J}(a, b) = +\infty$. We verify two properties of $\mathcal{J}$.
(P_1) $\mathcal{J}(\Phi_t(u), \Phi_t(v))$ is monotone nondecreasing in t.
It suffices to establish $\mathcal{J}(\Phi_t(u), \Phi_t(v)) \geq \mathcal{J}(u, v)$. For $s \leq \mathcal{J}(u, v)$, $\Phi_s(u) \leq v$, so $\Phi_s\Phi_t(u) \leq \Phi_t\Phi_s(u) \leq \Phi_t(v)$. Hence
$$\mathcal{J}(\Phi_t(u), \Phi_t(v)) \geq s \text{ for all } s \leq \mathcal{J}(u, v).$$
(P_2) If $u_k \leq v_k$, $u_k \in \overline{O(x)}$, $v_k \in \overline{O(y)}$ and $u_k \to u$, $v_k \to v$ then
$$\limsup_{k \to \infty} \mathcal{J}(u_k, v_k) \leq \mathcal{J}(u, v).$$

We may suppose that $\mathcal{J}(u, v) < \infty$ and $\limsup_{k \to \infty} \mathcal{J}(u_k, v_k) - \epsilon > \mathcal{J}(u, v)$ for some $\epsilon > 0$. Let $k_i \to \infty$ be such that $\lim_{i \to \infty} \mathcal{J}(u_{k_i}, v_{k_i}) = \limsup_{k \to \infty} \mathcal{J}(u_k, v_k)$. Then $\mathcal{J}(u, v) + \epsilon < \mathcal{J}(u_{k_i}, v_{k_i})$ for all large i. From the definition of $\mathcal{J}$, it follows that $\Phi_s(u_{k_i}) \leq v_{k_i}$ for $0 \leq s \leq \mathcal{J}(u, v) + \epsilon$ and all large i. Letting $i \to \infty$, we obtain $\Phi_s(u) \leq v$ for $0 \leq s \leq \mathcal{J}(u, v) + \epsilon$, a contradiction to the definition of $\mathcal{J}(u, v)$. Thus (P_2) holds.

It follows from (P_1) that $\alpha = \lim_{t \to \infty} \mathcal{J}(\Phi_t(x), \Phi_t(y))$ exists in $[0, \infty]$. By (P_2), $\mathcal{J} \equiv \mathcal{J}(a, b) \geq \alpha$. Suppose $\mathcal{J} < \infty$. For $0 \leq r \leq \mathcal{J}$, $\Phi_r(a) \leq b$. Actually, $\Phi_r(a) < b$, $0 \leq r \leq \mathcal{J}$ since otherwise $a, b \in \omega(x)$ by invariance of $\omega(x)$ and $a < b$ gives a contradiction to the Nonordering of Limit Sets. Let $K = \{\Phi_r(a) : 0 \leq r \leq \mathcal{J}\}$ so K is compact and $K < b$. By Lemma 3.3, there exists $t_1, \epsilon > 0$ and open sets U and V with $K \subset U$ and $b \in V$ such that $\Phi_{t_1+\delta}(U) \leq \Phi_{t_1}(V)$, $0 \leq \delta \leq \epsilon$. It follows that $\Phi_{t_k}(y) \in V$ for $k \geq k_0$ for some integer k_0. Since $\Phi_{t_k}(x) \to a$ as $k \to \infty$, $\Phi_r(\Phi_{t_k}(x)) \to \Phi_r(a)$ uniformly in $r \in [0, \mathcal{J}]$ as $k \to \infty$. Consequently, there exists k_1 such that $\Phi_r(\Phi_{t_k}(x)) \in U$ for all $k \geq k_1$ $0 \leq r \leq \mathcal{J}$. Then
$$\Phi_{t_1+\delta}\Phi_r\Phi_{t_k}(x) \leq \Phi_{t_1}\Phi_{t_k}(y)$$
for $k \geq k_2 = \max\{k_0, k_1\}$, $0 \leq r \leq \mathcal{J}$, $0 \leq \delta \leq \epsilon$. On rearranging and by monotonicity,
$$\Phi_{r+\delta}\Phi_{t+t_k}(x) \leq \Phi_{t+t_k}(y),$$
for $t \geq t_1$, $k \geq k_2$, $0 \leq r \leq \mathcal{J}$, $0 \leq \delta \leq \epsilon$. It follows that $\mathcal{J}(\Phi_{t+t_k}(x), \Phi_{t+t_k}(y)) \geq \mathcal{J} + \epsilon$, for $k \geq k_2$. Letting $k \to \infty$, we obtain $\alpha \geq \mathcal{J} + \epsilon$. But $\mathcal{J} \geq \alpha$ and this provides a contradiction. Hence $\mathcal{J} = \infty$ and the conclusion of the Proposition follows. □

PROPOSITION 3.5. *Absorption Principle.* Let $u, v \in X$. If there exists $x \in \omega(u)$ such that $x < \omega(v)$, then $\omega(u) < \omega(v)$. Similarly, if there exists $x \in \omega(u)$ such that $\omega(v) < x$, then $\omega(v) < \omega(u)$.

PROOF. Apply Lemma 3.3 above to obtain open neighborhoods U of x and V of $\omega(v)$ and $t_0 > 0$ such that $\Phi_{t_0}(U) \leq \Phi_{t_0}(V)$. Therefore, $\Phi_{t_0}(U) \leq \omega(v)$ since $\omega(v)$ is invariant. As $x \in \omega(u)$, there exists $t_1 > 0$ such that $\Phi_{t_1}(u) \in U$. Hence $\Phi_{t_0+t_1}(u) \leq \omega(v)$ and monotonicity implies that $\Phi_{t_0+t_1+s}(u) \leq \omega(v)$ for all $s \geq 0$. This implies that $\omega(u) \leq \omega(v)$. If $z \in \omega(u) \cap \omega(v)$ then, as $z \leq \omega(v)$ and $\omega(u) \leq z$, Corollary 2.4 implies that $\omega(u) = \omega(v) = z$. But this is impossible since $x < \omega(v)$ and $x \in \omega(u)$ so we conclude that $\omega(u) < \omega(v)$. □

PROPOSITION 3.6. *Limit Set Separation Principle.* Let x, y satisfy $x < y$. If $t_k \to \infty$, $\Phi_{t_k}(x) \to a$, $\Phi_{t_k}(y) \to b$ as $k \to \infty$ and $a < b$ then $\omega(x) < \omega(y)$.

PROOF. By Proposition 3.4, $O(a) \leq b$ and therefore $\omega(a) \leq b$. If $b \in \omega(a)$ then Corollary 2.4 implies that $\omega(a) = b \in E$. Applying the Absorption Principle with $u = x, v = a, x = a$, we have $a \in \omega(x)$, $a < \omega(a) = b$ which implies that $\omega(x) < \omega(a)$. This is impossible as $\omega(a) \subset \omega(x)$. Consequently, $\omega(a) < b$. By the Absorption Principle again ($u = a, v = y$) $\omega(a) < \omega(y)$. Since every $z \in \omega(a)$ belongs to $\omega(x)$ as well, the Absorption Principle gives $\omega(x) < \omega(y)$. □

We can now prove one of the fundamental results in the theory.

THEOREM 3.7. *Limit Set Dichotomy.* If $x < y$ then either
(a) $\omega(x) < \omega(y)$, or
(b) $\omega(x) = \omega(y) \subset E$.
If case (b) holds and $t_k \to \infty$ then $\Phi_{t_k}(x) \to p$ if and only if $\Phi_{t_k}(y) \to p$.

PROOF. If $\omega(x) = \omega(y)$ then $\omega(x) \subset E$ by the Intersection Principle. The remark following the proof of the Intersection Principle establishes the final assertion. If $\omega(x) \neq \omega(y)$ then we may assume that there exists $q \in \omega(y) \setminus \omega(x)$, the other case is treated similarly. There exists $t_k \to \infty$ such that $\Phi_{t_k}(y) \to q$. By passing to a subsequence if necessary, we can assume that $\Phi_{t_k}(x) \to p \in \omega(x)$. Monotonicity implies $p \leq q$ and, in fact, $p < q$ since $q \notin \omega(x)$. By the Limit Set Separation Principle, $\omega(x) < \omega(y)$. □

4. Quasiconvergence is Generic

The main result of this chapter says that the typical orbit of an SOP semiflow converges to the set E. Our proof requires an additional compactness assumption which we introduce below. First, we need a definition.

If $x \in X$, we say that *x can be approximated from below (above) in X* if there exists a sequence $\{x_n\}$ in X satisfying $x_n < x_{n+1} < x$ ($x < x_{n+1} < x_n$) for $n \geq 1$ and $x_n \to x$ as $n \to \infty$.

(C) For each $x_0 \in X$, $O(x_0)$ has compact closure in X. Furthermore, if $\{x_n\}_{n \geq 1}$ approximates x_0 from below or from above then $\cup_{n \geq 0} \omega(x_n)$ has compact closure contained in X.

(C) is a relatively weak compactness requirement. It is satisfied if (1): every bounded (or even just compact) set $B \subset X$ has a bounded orbit $O(B) = \cup_{x \in B} O(x)$, and (2): Φ_t is *conditionally completely continuous* for some $t_0 > 0$. Recall that Φ_t is (conditionally)·completely continuous if $\Phi_t(B)$ has compact closure in X whenever B is a bounded subset of X (and $\Phi_t(B)$ is a bounded set). If these two conditions are met and if $\{x_n\}_{n\geq 1}$ approximates x_0 from below then $\{x_n\}_{n\geq 0}$ is compact and therefore $\cup_{n\geq 0} O(x_n)$ is bounded. The conditional compactness of Φ_{t_0} implies that $\overline{\Phi_{t_0}(\cup_{n\geq 0} O(x_n))}$ is compact in X. It clearly contains $\cup_{n\geq 0}\omega(x_n)$ and therefore the latter has compact closure in X.

(C) is assumed to hold throughout the remainder of this chapter. The key to our proof that a generic point of X is a quasiconvergent point is the following result.

THEOREM 4.1. *Sequential Limit Set Trichotomy. Let $x_0 \in X$ have the property that it can be approximated from below in X by a sequence $\tilde{x}_n$. Then there exists a subsequence x_n of $\tilde{x}_n$ such that $x_n < x_{n+1} < x_0$, $n \geq 1$, with $x_n \to x_0$ satisfying one of the following.*
(a) *There exists $u_0 \in E$ such that*
$$\omega(x_n) < \omega(x_{n+1}) < u_0 = \omega(x_0) \quad , \quad n \geq 1$$
and
$$\lim_{n\to\infty} dist\,(\omega(x_n), u_0) = 0.$$
(b) *There exists $u_0 \in E$ such that*
$$\omega(x_n) = u_0 < \omega(x_0) \quad , \quad n \geq 1.$$
If $u \in E$ and $u < \omega(x_0)$ then $u \leq u_0$.
(c) $\omega(x_n) = \omega(x_0) \subset E$ *for* $n \geq 1$.

PROOF. Let $\tilde{x}_n$ be any sequence satisfying $\tilde{x}_n < \tilde{x}_{n+1} < x_0$, $n \geq 1$ and $\tilde{x}_n \to x_0$. By the Limit Set Dichotomy either there exists a positive integer N such that $\omega(\tilde{x}_n) = \omega(\tilde{x}_m)$ for all m, n larger than N or there is a subsequence $\tilde{x}_{n_i}$ such that $\omega(\tilde{x}_{n_i}) < \omega(\tilde{x}_{n_{i+1}})$ for all i. By passing to this subsequence or renumbering the sequence, we may assume that either $\omega(x_n) = \omega(x_m)$ for all n, m or $\omega(x_n) < \omega(x_{n+1})$ for $n \geq 1$, where x_n is the appropriate subsequence of $\tilde{x}_n$.

Suppose that the latter is the case. Then $\omega(x_n) < \omega(x_0)$ for all n. For if $\omega(x_n) = \omega(x_0)$ for some $n = n_0$ then $\omega(x_n) = \omega(x_0)$ for $n \geq n_0$, a contradiction to $\omega(x_n) < \omega(x_{n+1})$. Let $\Omega = \{y : y = \lim y_n, y_n \in \omega(x_n)\} \subset \overline{\cup_{n\geq 1}\omega(x_n)}$. By (C), $\cup_{n\geq 1}\omega(x_n)$ has compact closure in X and since $\{y_n\}$ is monotone and contained in a compact set, it converges. Hence Ω is nonempty and compact. Suppose y and u belong to Ω so that $y_n \to y$, $u_n \to u$ where $y_n, u_n \in \omega(x_n)$. Since $y_n < u_{n+1}$ and $u_n < y_{n+1}$ holds for all n, we obtain $y \leq u$ and $u \leq y$ so $u = y$. Thus Ω is a singleton, $\Omega = \{u_0\}$. Furthermore, Ω is positively invariant since each $\omega(x_n)$ is invariant. Thus $u_0 \in E$. It follows immediately from the definition of Ω and the fact that $\cup_{n\geq 1}\omega(x_n)$ has compact closure in X that $\lim_{n\to\infty} dist\,(\omega(x_n), u_0) = 0$. Finally, $\omega(x_n) < \omega(x_0)$ for all n implies that $u_0 \leq \omega(x_0)$. If $u_0 \in \omega(x_0)$ then $\omega(x_0) = u_0$ by Corollary 2.4. This is just (a) above. If $u_0 < \omega(x_0)$ then choose a neighborhood W of $\omega(x_0)$ and $t_0 \geq 0$ such that $u_0 \leq \Phi_t(W)$ for $t \geq t_0$ (by Lemma 3.3). Now there exists $t_1 > 0$ such that $\Phi_{t_1}(x_0) \in W$ and by continuity of Φ_{t_1}, there is an integer n such that $\Phi_{t_1}(x_n) \in W$. It follows that $u_0 \leq \Phi_t(x_n)$ for $t \geq t_0 + t_1$. But $u_0 \in E$ so necessarily, $\omega(x_n) \geq u_0$. On the other hand, $\omega(x_n) < \omega(x_{n+1}) \leq u_0$ holds for every n so $\omega(x_n) < u_0$ holds for every n. This contradiction shows that $\omega(x_0) = u_0$. Thus (a) holds if $\omega(x_n) < \omega(x_{n+1})$, $n \geq 1$.

Suppose now that $\omega(x_n) = \omega(x_1)$, $n \geq 1$. Since $x_n < x_0$, the Limit Set Dichotomy implies that either $\omega(x_n) = \omega(x_1) < \omega(x_0)$ or $\omega(x_n) = \omega(x_0)$ for $n \geq 1$. The latter is precisely case (c). Suppose $\omega(x_1) < \omega(x_0)$. Let $u_0 \in \omega(x_1) = \omega(x_n) \subset E$ so $u_0 < \omega(x_0)$. By the Lemma 3.3 there exists an open set W containing $\omega(x_0)$ and $t_0 \geq 0$ such that $u_0 \leq \Phi_t(W)$ for $t \geq t_0$. Now, arguing exactly as in the previous paragraph, we obtain $u_0 \leq \Phi_t(x_n)$ for some n and all large t. This yields $\omega(x_n) \geq u_0$ and since $u_0 \in \omega(x_n)$ it follows that $\omega(x_n) = u_0$ by Corollary 2.4. Thus $\omega(x_1) = \omega(x_n) = u_0$ as asserted in case (b). Finally, if $u \in E$ and $u < \omega(x_0)$, we may argue exactly as above with u_0, that $u_0 = \omega(x_n) \geq u$ for all large n, establishing that $u_0 \geq u$. $\square$

Clearly, an analogous result holds if x_0 can be approximated from above in X. Although we do not state such a result formally, it will be used below.

Of the three alternatives for the point x_0 described in the Sequential Limit Set Trichotomy, (a) implies that it is a convergent point, (c) implies that it is a quasiconvergent point and (b) implies that it belongs to the closure of the set of convergent points. In our next result we summarize some information which follows from the proof of the Sequential Limit Set Trichotomy and that will allow us to strengthen the last assertion concerning case (b). Only those assertions concerning case (b) will be used in this chapter.

COROLLARY 4.2. *Assume that the hypotheses of the Sequential Limit Set Trichotomy hold. In cases* (a), (b) *and* (c) *we have, in addition, the following:*
(a) *For any n there exists a neighborhood U_n of x_0 and $t_n \geq 0$ such that*

$$\Phi_t(x_n) \leq \Phi_t(U_n) \quad \text{for} \quad t \geq t_n.$$

(b)
  (i) *There exists a neighborhood O of u_0, $t_0, t_1 \geq 0$, and n such that*

$$\Phi_t(O) \leq \Phi_{t+t_1}(x_n) \quad \text{for} \quad t \geq t_0.$$

  (ii) *There is a neighborhood U of x_0 with the following property: for each $x \in U$, $x < x_0$, there exists a neighborhood $V = V_x$ of x in U, an integer $N = N_x$ and $T = T_x > 0$ such that*

$$u_0 \leq \Phi_t(V) \leq \Phi_t(x_N) \quad \text{for } t \geq T.$$

(c) *There is a neighborhood U of x_0 with the following property: for each $x \in U$, $x < x_0$, there exists a neighborhood $V = V_x$ of x in U and $T = T_x > 0$ such that*

$$\Phi_t(x_1) \leq \Phi_t(V) \leq \Phi_t(x_0) \quad \text{for } t \geq T.$$

Moreover,

$$d(\Phi_t(x_0), \Phi_t(x_1)) \to 0 \quad \text{as } t \to \infty.$$

PROOF. For (a), fix n. Since $\omega(x_n) < \omega(x_0)$, Lemma 3.3 implies the existence of neighborhoods $W_1 \supset \omega(x_n)$ and $W_2 \supset \omega(x_0)$ and $t_0 \geq 0$ such that $\Phi_t(W_1) \leq \Phi_t(W_2)$ for $t \geq t_0$. There exists $t_1 > 0$ such that $\Phi_{t_1}(x_n) \in W_1$ and $\Phi_{t_1}(x_0) \in W_2$. By continuity of Φ_{t_1}, there is a neighborhood U_n of x_0 such that $\Phi_{t_1}(U_n) \subset W_2$. Hence $\Phi_{t+t_1}(x_n) \leq \Phi_{t+t_1}(U_n)$ for $t \geq t_0$, so (a) holds with $t_n = t_0 + t_1$.

Now consider (b). Since $u_0 < \omega(x_0)$, by Lemma 3.3, there exists a neighborhood W of $\omega(x_0)$, a neighborhood O of u_0 and $t_0 \geq 0$ such that $\Phi_t(O) \leq \Phi_t(W)$

for $t \geq t_0$. There exists $t_1 > 0$ such that $\Phi_t(x_0) \in W$ for $t \geq t_1$. By continuity of Φ_{t_1}, there exists n such that $\Phi_{t_1}(x_n) \in W$. It follows that

$$\Phi_t(O) \leq \Phi_{t+t_1}(x_n) \quad , \quad t \geq t_0.$$

Choose a neighborhood U of x_0 such that $\Phi_{t_1}(U) \subset W$ and let $x \in U$ satisfy $x < x_0$. Then there exist neighborhoods V of x, $V \subset U$, $\mathcal{N}$ of x_0 and $t_2 \geq 0$ such that $\Phi_t(V) \leq \Phi_t(\mathcal{N})$ for $t \geq t_2$. Choosing N such that $x_N \in \mathcal{N}$, we have $\Phi_t(V) \leq \Phi_t(x_N)$ for $t \geq t_2$. From the paragraph above, we have that

$$u_0 = \Phi_t(u_0) \in \Phi_t(O) \leq \Phi_t(W), \quad t \geq t_0.$$

As $\Phi_{t_1}(V) \subset \Phi_{t_1}(U) \subset W$, it follows that

$$u_0 \in \Phi_t(O) \leq \Phi_t(\Phi_{t_1}(V)) = \Phi_{t+t_1}(V) \quad , \quad t \geq t_0,$$

hence

$$u_0 \leq \Phi_t(V) \leq \Phi_t(x_N) \quad , \quad t \geq t_0 + t_1 + t_2.$$

This establishes (b) with $T = t_0 + t_1 + t_2$.

Now consider (c). Since $x_1 < x_0$, there exists a neighborhood U of x_0 and $t_3 \geq 0$ such that $\Phi_t(x_1) \leq \Phi_t(U)$ for $t \geq t_3$. Hence, if $x \in U$ and $x < x_0$ then there is a neighborhood V of x, $V \subset U$, and $t_4 \geq 0$ such that $\Phi_t(V) \leq \Phi_t(x_0)$ for $t \geq t_4$. Since $V \subset U$, it follows that

$$\Phi_t(x_1) \leq \Phi_t(V) \leq \Phi_t(x_0) \quad \text{for } t \geq t_3 + t_4.$$

If $d(\Phi_t(x_0), \Phi_t(x_1)) \to 0$ as $t \to \infty$ does not hold, then there exists $\epsilon > 0$, $t_n \to \infty$ such that $d(\Phi_{t_n}(x_0), \Phi_{t_n}(x_1)) \geq \epsilon$ for $n \geq 1$. Without loss of generality, we can assume $\Phi_{t_n}(x_0) \to u$, $\Phi_{t_n}(x_1) \to v$ where $u, v \in \omega(x_0)$. But $u \geq v$ and the Nonordering of Limit Sets implies that $u = v$. This contradiction completes the proof of the Corollary. $\square$

REMARK 4.1. Corollary 4.2 contains stability information which will be exploited more fully in a later chapter under the additional hypothesis that X is normally ordered. Of the following remarks, only (b) (ii) will be used in the proof of the main result below.
(b)(i): Observe that if $x \in O$ and $x > u_0$ then $u_0 < \Phi_t(x) \leq \Phi_{t+t_1}(x_n)$ for $t \geq t_0$. Since $\Phi_t(x_n) \to u_0$ as $t \to \infty$, $\omega(x) = u_0$. Thus, $\omega(x) = u_0$ for all $x \in O$ with $x \geq u_0$.
(b)(ii): Similarly $\Phi_t(x_N) \to u_0$ as $t \to \infty$, so $\omega(v) = u_0$ for all $v \in V$. In particular, $\omega(x) = u_0$ and $x \in \text{Int} C$ for all $x \in U$ with $x < x_0$. Therefore, $x_0 \in \overline{\text{Int} C}$.
(c) Similar arguments imply that $\omega(x) = \omega(x_0)$ for all $x \in U$ with $x < x_0$.

The main result of this chapter is the following. It says that a generic point of X is a quasiconvergent point.

THEOREM 4.3. *Suppose each point of X can be approximated either from above or from below in X. Then $X = \text{Int} Q \cup \overline{\text{Int } C}$. In particular, $\text{Int} Q$ is dense in X.*

PROOF. Suppose $x_0 \in X \setminus \text{Int } Q$. Then there exists a sequence $y_n \in X \setminus Q$ such that $y_n \to x_0$. By passing to a subsequence if necessary, we can assume that either y_n can be approximated from below in X for each n or y_n can be approximated from above in X for each n. We consider only the former case as the latter case is similar (using the Sequential Limit Set Trichotomy and Corollary 4.2 where x_0 is approximated from above). Each y_n is the limit of a sequence $x_m^n \to y_n$, $x_m^n < x_{m+1}^n < y_n$. For each y_n, case (b) of the Limit Set Trichotomy must hold since $y_n \notin Q$. By (b) (ii) of Corollary 4.2 and Remark 4.1, it follows that $y_n \in \overline{\text{Int } C}$ for each n because $x_m^n \in \text{Int } C$ for all large m. Hence $x_0 \in \overline{\text{Int } C}$. □

REMARK 4.2. There are two conditions which imply that $Q = C$ that are worth mentioning. If E has no points of accumulation in X then $Q = C$. This is because $\omega(x)$ is connected so if it contains more than one point then each of its points is a point of accumulation of $\omega(x)$. The condition that E has no points of accumulation in X is typically satisfied in the case of ordinary differential equations but is difficult to verify for most infinite dimensional systems. If E is *totally ordered*, that is, $x \leq y$ or $y \leq x$ holds for any pair of points of E, then $Q = C$. This follows from Nonordering of Limit Sets. The condition that E is totally ordered holds for scalar delay differential equations as we note in a later chapter.

It is convenient to have the following immediate consequence of the Limit Set Trichotomy for reference in the next chapter. It describes the possibilities for the omega limit sets in the case that a point can be approximated from above and from below in X.

PROPOSITION 4.4. *Let X be an ordered metric space and Φ_t be a strongly order preserving semiflow on X. Let $x_0 \in X$ be such that it can be approximated from above in X and from below in X. Then there exists sequences x_n and z_n in X satisfying $x_n \to x_0$, $z_n \to x_0$, $x_n < x_{n+1} < x_0 < z_{n+1} < z_n$, $n \geq 1$, and one of the following holds:*
(a) *There exists $u_0 \in E$ such that, for $n \geq 1$,*

$$\omega(x_n) < \omega(x_{n+1}) < \omega(x_0) = u_0 < \omega(z_{n+1}) < \omega(z_n)$$

and

$$\lim_{n \to \infty} \text{dist}(\omega(x_n), u_0) = \lim_{n \to \infty} \text{dist}(\omega(z_n), u_0) = 0.$$

(b) *There exists $u_0, v_0 \in E$ such that, for $n \geq 1$, either*
   (i) $\omega(x_n) < \omega(x_{n+1}) < \omega(x_0) = u_0 < v_0 = \omega(z_n)$,

$$\lim_{n \to \infty} \text{dist}(\omega(x_n), u_0) = 0$$

and whenever $v \in E$, $v > u_0$ then $v \geq v_0$,
or
   (ii) $\omega(x_n) = u_0 < v_0 = \omega(x_0) < \omega(z_{n+1}) < \omega(z_n)$

$$\lim_{n \to \infty} \text{dist}(\omega(z_n), v_0) = 0$$

and whenever $u \in E$ and $u < v_0$ then $u \leq u_0$.
(c) *There exists $u_0 \in E$ such that, for $n \geq 1$, either*
   (i) $\omega(x_n) < \omega(x_{n+1}) < \omega(x_0) = u_0 = \omega(z_n)$,

$$\lim_{n \to \infty} \text{dist}(\omega(x_n), u_0) = 0$$

or

(ii) $\omega(x_n) = u_0 = \omega(x_0) < \omega(z_{n+1}) < \omega(z_n)$ and
$$\lim_{n\to\infty} dist\,(\omega(z_n), u_0) = 0.$$

(d) There exist equilibria u_0 and v_0 such that, for $n \geq 1$,
$$\omega(x_n) = u_0 < \omega(x_0) < v_0 = \omega(z_n).$$

If $u \in E$ and $u < \omega(x_0)$ then $u \leq u_0$. If $v \in E$ and $\omega(x_0) < v$ then $v \geq v_0$.

(e) There exists $u_0 \in E$ such that, for $n \geq 1$, either

(i) $\omega(x_n) = u_0 < \omega(x_0) = \omega(z_n) \subset E$ and, whenever $u \in E$ satisfies $u < \omega(x_0)$, then $u \leq u_0$

or

(ii) $\omega(z_n) = u_0 > \omega(x_0) = \omega(x_n) \subset E$ and, whenever $u \in E$ satisfies $u > \omega(x_0)$, then $u \geq u_0$.

(f) For $n \geq 1$, $\omega(x_n) = \omega(x_0) = \omega(z_n) \subseteq E$.

The sequence $x_n(z_n)$ can be chosen to be a subsequence of any sequence $\tilde{x}_n(\tilde{z}_n)$ approximating x_0 from below (above) in X.

5. Remarks and Discussion

The results of sections 2 and 3 are all due originally to Hirsch(1988b). An exception is Theorem 2.2 which was first established by Hadeler and Glas(1983) for ordinary differential equations and later more generally by Hirsch (1984,1988b).

Hirsch assumes that the semiflow is strongly monotone and this requires that the partial order relation have nonempty interior. In applications, the cone generating the partial order relation must have nonempty interior. Our treatment follows Smith and Thieme(1990a) who show that Hirsch's results are valid under the weaker assumption that the semiflow is strongly order preserving. Matano(1986,1987) introduced the strong order preserving hypothesis in his investigations of parabolic partial differential equations. He pointed out that the SOP property allows more flexibility in the choice of state space X than the strong monotonicity assumption. This will be born out in the applications of the last three chapters.

Smith and Thieme(1990a) introduce the compactness hypothesis (C) and obtain the Sequential Limit Set Trichotomy. This tool seems to streamline the arguments leading to Theorem 4.3 and allows a stronger conclusion than the comparable result of Hirsch(1988b), Theorem 7.5. The latter result concludes that the set of quasiconvergent points is a residual subset of X (it contains a countable intersection of open and dense subsets).

It is not necessary to assume that the semiflow is globally defined, that is, that all solutions are defined for $t \geq 0$. See Hirsch(1988b) and Smith and Thieme(1991a).

Takáč(1991) introduces the idea of ω-stability for a strongly monotone semiflow and considers the domain of attraction of an omega limit set of a generic order ω-stable point.

CHAPTER 2

Stability and Convergence

0. Introduction

In the previous chapter we showed that the generic orbit of a strongly monotone dynamical system approaches the set of equilibria. Here, we focus on obtaining additional conditions which guarantee that the orbit actually approaches a single equilibrium. Stability notions play a considerable role. The principal result of the chapter, Theorem 4.7, gives sufficient conditions for the generic orbit of a strongly order preserving semiflow to converge to equilibrium. This improvement of Theorem 1.4.3 requires that the semiflow is continuously differentiable in the state variable. Making essential use of the Krein-Rutman Theorem, a sharper version of the Limit Set Dichotomy is obtained which leads to the improvement.

We are also interested in conditions which imply that all solutions converge to equilibrium. One of the more important of these results is the Order Interval Trichotomy. It describes the possible asymptotic behavior of a strongly order preserving, completely continuous, semiflow on an order interval bounded by two equilibria. Either there is a third equilibrium in the interval or all non-equilibrium solutions approach the same boundary equilibrium. In a different direction, we show that if there is only one equilibrium, then it is globally attracting. If there do not exist three ordered equilibria $a < b < c$, then all solutions are shown to approach an equilibrium.

Finally, we introduce the notion of sub and super-equilibria. These are points belonging to monotone solutions. The existence of monotone heteroclinic trajectories, i.e., full orbits connecting two order related equilibria, are shown to exist.

1. Stability

Let X be a subset of the Banach space Y with cone Y_+. Y_+ generates a partial order on Y and, by restriction, on X as well. The norm of $y \in Y$ is denoted by $|y|$. If $u, v \in Y$ satisfy $u < v$ then $[u, v] \equiv \{y \in Y : u \leq y \leq v\}$ is called the *order interval* generated by u, v. Until further notice, we assume that the cone Y_+ is *normal*. This means that there exists a constant $k > 0$ such that whenever $0 \leq x \leq y$ then $|x| \leq k|y|$. Equivalently, all order intervals are bounded (see Amann(1976)). $\Phi_t(x)$ is assumed to be an SOP semiflow on X satisfying the compactness hypothesis (C).

If $x \in X$ then x is a *stable point* if for every $\epsilon > 0$ there exists $\delta > 0$ such that $|\Phi_t(x) - \Phi_t(y)| < \epsilon$ for $t \geq 0$ whenever $y \in X$ and $|x - y| < \delta$. Let S be the subset of stable points of X. A point x is an *asymptotically stable point* if there is a neighborhood V of x with the property that for every $\epsilon > 0$ there exists $t_\epsilon > 0$ such that $|\Phi_t(x) - \Phi_t(y)| < \epsilon$ if $t \geq t_\epsilon$ and $y \in V$. Let A denote the set of all asymptotically stable points. A point $x \in X$ is *asymptotically stable from above* (*below*) if there is a neighborhood V of x with the property that for every $\epsilon > 0$ there exists $t_\epsilon > 0$ such that $|\Phi_t(x) - \Phi_t(y)| < \epsilon$ if $t \geq t_\epsilon$, $y \in V$ and $y > x$ ($y < x$).

A stable point x_0 has the property that nearby points have omega limit sets which are nearby $\omega(x_0)$. In fact it is easily seen that the map $x \to \omega(x)$ is continuous at x_0 in the Hausdorff metric if $x_0 \in S$. If x_0 is an asymptotically stable point then evidently, $\lim_{t \to \infty} |\Phi_t(x_0) - \Phi_t(x)| = 0$ uniformly in $x \in V$. Consequently, $\omega(x) = \omega(x_0)$ for all $x \in V$.

PROPOSITION 1.1. *A is an open subset of X and $A \subset S$. If every point of X can be approximated either from above or from below in X then $S \subset Q$.*

PROOF. If $x \in A$ then every point of the neighborhood V in the definition of asymptotically stable point belongs to A so A is open. The containment $A \subset S$ follows from the continuity of Φ. If $x \in S$ then nearby points have nearby limit sets. It follows that only alternatives (a) and (c) of the Sequential Limit Set Trichotomy are possible. Thus $x \in Q$. □

The normality of the cone Y_+ plays a crucial role in the following.

REMARK 1.1. In (b) and (c) of Corollary 1.4.2, each $x \in U$ with $x < x_0$ belongs to A. Indeed, the neighborhood V of x, in (b)(ii) of Corollary 1.4.2, can be taken for the neighborhood V in the definition of asymptotically stable point. For if $y \in V$ then

$$|\Phi_t(x) - \Phi_t(y)| \leq |\Phi_t(x) - u_0| + |\Phi_t(y) - u_0| \leq 2k\,|\Phi_t(x_N) - u_0|, \quad t \geq T.$$

Similarly, in case (c), if $y \in V$ then

$$|\Phi_t(x) - \Phi_t(y)| \leq |\Phi_t(x) - \Phi_t(x_0)| + |\Phi_t(x_0) - \Phi_t(y)| \leq 2k|\Phi_t(x_0) - \Phi_t(x_1)|, \quad t \geq T.$$

Since the term on the right hand side of each inequality tends to zero, it follows that $x \in A$. Similarly, the equilibrium u_0 in (b)(i) is asymptotically stable from above since for every $\epsilon > 0$ there exists $t_\epsilon > 0$ such that if $x \in O$ and $x > u_0$ then $|\Phi_t(x) - u_0| < \epsilon$ for $t > t_\epsilon$.

The next result says that a generic point of X is either an asymptotically stable point, which is quasiconvergent by Proposition 1.1, or belongs to the interior of the set of convergent points.

THEOREM 1.2. *Assume that each $x \in X$ can be approximated from above or from below in X. Then $A \cup \operatorname{Int} C$ is dense in X.*

PROOF. If $A \cup \operatorname{Int} C$ is not dense in X there exists an open set U in X such that $U \cap A = \emptyset = U \cap \operatorname{Int} C$. Let $x \in U$ and assume that x can be approximated from below. Then there is a sequence x_n such that $x_n < x_{n+1} < x$, $x_n \to x$ and one of the alternatives (a),(b) or (c) of the Sequential Limit Set Trichotomy and Corollary 1.4.2 holds. We can assume $x_n \in U$ for all n. As $U \cap A = \phi$, only case (a) can hold (see Remark 1.1). Hence x is convergent. Since $x \in U$ was arbitrary, $U \subset C$. So $U \subset \operatorname{Int} C$ in contradiction to our assumption. □

2. The Order Interval Trichotomy

The following result is an important tool which will be used repeatedly in the remainder of this chapter. It gives sufficient conditions for the existence of three ordered equilibria. The proof uses the fixed point index (see Amann(1976)).

PROPOSITION 2.1. *Let Φ be SOP on $X = [u, v]$ where $u < v$ and $u, v \in E$. Suppose that u and v are asymptotically stable in X and that $\overline{\Phi_t([u,v])}$ is compact for each $t > 0$. Then there exists $w \in [u, v] \cap E$ such that $w \neq u, v$.*

PROOF. Fix $t_0 > 0$ and consider the map Φ_{t_0} restricted to $[u, v]$. We use the fixed point index to establish the existence of a fixed point for Φ_{t_0} in X different from u and v. Hereafter, all topological notions are to be understood relative to X.

Since u is asymptotically stable in X, there exists $r > 0$ such that each point x in the closure of $B(u) = \{x \in X : |u - x| < r\}$ satisfies $\omega(x) = u$. Since $v \in A$, by taking r smaller if necessary, each point x in the closure of $B(v) = \{x \in X : |x - v| < r\}$ satisfies $\omega(x) = v$ and $B(u) \cap B(v) = \emptyset$. We will establish that the fixed point indices below are defined and satisfy

$$i(\Phi_{t_0}, B(u), X) = i(\Phi_{t_0}, B(v), X) = i(\Phi_{t_0}, X, X) = +1.$$

By the additivity property and the solution property of the fixed point index, this implies that Φ_{t_0} has a fixed point w in $X \setminus (B(u) \cup B(v))$. Define the homotopy $F : [0, 1] \times \overline{B(u)} \to X$ by $F(\lambda, x) = \lambda u + (1 - \lambda)\Phi_{t_0}(x)$. If $F(\lambda, x) = x$ then $x - \Phi_{t_0}(x) = \lambda(u - \Phi_{t_0}(x)) \leq 0$ because $[u, v]$ is invariant under Φ. Equality can only hold if $x = u$ because u is the only fixed point of Φ_{t_0} in $\overline{B(u)}$ (recall that $\omega(x) = u$). Inequality implies $\Phi_{t_0}(x) > x$ which, by the Convergence Criterion, gives that $\Phi_t(x)$ converges to an equilibrium larger than x. But this is impossible for any $x \in \overline{B(u)}$ because $\omega(x) = \{u\}$. We have shown that the fixed point set of $F(\lambda, \cdot)$ in $\overline{B(u)}$ is precisely $\{u\}$. Therefore, the homotopy property of the fixed point index implies that

$$i(\Phi_{t_0}, B(u), X) = i(F(1, \cdot), B(u), X) = +1.$$

The latter equality holds because the index of a constant mapping with value in $B(u)$ is $+1$. Similar arguments show that $i(\Phi_{t_0}, B(v), X) = +1$. The homotopy F above can be considered on $[0, 1] \times X$. Since F maps the product space into X and has no fixed points on the boundary of X (**relative to X**) since this boundary is empty, it follows that $i(\Phi_{t_0}, X, X) = +1$. Thus there exists $w \in X \setminus (B(u) \cup B(v))$ such that $\Phi_{t_0}(w) = w$. As $t_0 > 0$ was arbitrary, we obtain fixed points $w_n \in [u, v] \setminus (B(u) \cap B(v))$ of $\Phi_{2^{-n}}$ for all large n. The set $\{w_n\}$ is precompact in $[u, v]$ since each w_n is a fixed point of $\Phi_{2^{-j}}$ for all $n \geq j$ so $w_n \in \overline{\Phi_{2^{-j}}(X)}$ and $\overline{\Phi_{2^{-j}}(X)}$ is compact. A standard argument gives that a limit point of $\{w_n\}$ is an equilibrium of Φ in $[u, v] \setminus (B(u) \cup B(v))$: Let $w_{n_l} \to w$ for $l \to \infty$ and $t_l = 2^{-n_l}, \bar{w}_l = w_{n_l} \to w$. If $t > 0$, represent $t = m_l t_l + r_l$, $0 \leq r_l < t_l$, with a nonnegative integer m_l, $l = 1, 2, \cdots$. Then

$$\Phi_t(w) = \lim_l \Phi_t(\bar{w}_l) = \lim_l \Phi_{r_l} \Phi_{m_l t_l}(\bar{w}_l) = \lim_l \Phi_{r_l}(\bar{w}_l) = w.$$

The proof is complete. □

The following trichotomy is of fundamental importance in the theory as well as the applications.

THEOREM 2.2. *Let Φ be SOP on $X = [u, v]$ where $u < v$ and $u, v \in E$. If $\overline{\Phi_t([u,v])}$ is compact for each $t > 0$, then one of the following holds:*
(i) *there exists $w \in [u, v] \cap E$, $w \neq u, v$.*
(ii) *$\Phi_t(x) \to u$ for all $x \in [u, v] \setminus v$.*

(iii) $\Phi_t(x) \to v$ for all $x \in [u,v] \setminus u$.

PROOF. Suppose that (i) doesn't hold, that is, $X \cap E = \{u,v\}$. Evidently $Q = C$ in this case and if $x \in C$ then $\omega(x) = u$ or $\omega(x) = v$. We first show that $X = C$. For if $x_0 \in X \setminus C$ then $x_0 \neq u, v$ so it can be approximated from above by a sequence z_n of points on the segment joining x_0 and v and from below by a sequence x_n of points on the segment joining x_0 to u. Since $x_0 \notin Q$, case (d) of Proposition 1.4.4 holds:

$$\omega(x_n) = u < \omega(x_0) < v = \omega(z_n)$$

for $n \geq 1$. By Remark 1.1, u is asymptotically stable from above and (a similar argument implies that) v is asymptotically stable from below. But this just means that $u, v \in A$ and so Proposition 2.1 implies that there exists a third equilibrium in X. This contradiction implies that $X = C$.

Now suppose that there are elements $x, y \in [u,v] \setminus \{u,v\}$ with $\Phi_t(x) \to u$, $\Phi_t(y) \to v$, $t \to \infty$. As Φ is strongly order preserving, we find neighborhoods U, V of u, v and $t_0 > 0$ such that $\Phi_t(U) \leq \Phi_t(x)$, $\Phi_t(V) \geq \Phi_t(y)$ for $t \geq t_0$. As X is normally ordered, we have that

$$\Phi_t(U \cap [u,v]) \to u, \quad \Phi_t(V \cap [u,v]) \to v, \quad t \to \infty.$$

Therefore, $u, v \in A$ and Proposition 2.1 implies the existence of a third equilibrium. This contradiction completes the proof. □

3. Some Global Results

If there is only one equilibrium point in X then it must be globally attracting according to the next result. A common situation where this result may apply is the case that $X = Y_+$, the positive cone, and $e = 0$. However, the requirement that Φ is SOP on all of Y_+ is a severe one as we shall see later.

THEOREM 3.1. **Global Asymptotic Stability.** *Suppose that X contains exactly one equilibrium e and that every point of $X \setminus e$ can be approximated from above and from below in X. Then $\omega(x) = e$ for all $x \in X$.*

PROOF. If $x \in X \setminus e$ then only alternatives (a),(e) and (f) of Proposition 1.4.4 may hold since the others imply more than a single equilibrium. In particular, $x \in Q$ so $\omega(x) = e$. □

In some applications it may be known that there are at most two equilibria in X. For example, $X = Y_+$ and there is one nonzero equilibrium. As a special case of the next result, we may conclude that all solutions converge to one of these equilibria. The set X is said to be *order convex* if $[u,v] \subset X$ whenever $u, v \in X$ satisfy $u < v$.

THEOREM 3.2. *Suppose that X is an order convex subset of Y and Φ_t is completely continuous for $t > 0$. Assume that any point in $X \setminus E$ can be approximated both from above and from below. If X does not contain equilibria u, v, w such that $u < v < w$, then $X = Q$.*

PROOF. If $x \in X \setminus E$ then case (d) of Proposition 1.4.4 cannot hold since then there would be three ordered equilibria by Proposition 2.1 and remark 1.1. Therefore, $x \in Q$. □

A simple fact regarding the *basin of attraction* of an equilibrium e of a monotone semiflow should be mentioned here. Recall that the basin of attraction of e is the set of all $x \in X$ such that $\Phi_t(x) \to e$ as $t \to \infty$. If $x, y, z \in X$ satisfy $x \le y \le z$ and x, z belong to the basin of attraction of e then so does y by letting $t \to \infty$ in the comparison $\Phi_t(x) \le \Phi_t(y) \le \Phi_t(z)$. Here, we must assume that either Y_+ is normal or that $\overline{O(y)}$ is compact in X.

4. Generic Convergence to Equilibrium

The aim in this section is to show that Theorem 1.4.3 can be improved, by adding mild additional hypotheses, in order to conclude that most points are convergent points. The additional hypotheses include the assumptions that the semiflow is continuously differentiable and that the derivative is a compact, strongly positive linear operator. Therefore, it will be assumed that the cone Y_+ has nonempty interior, $\text{Int} Y_+$, in Y (see Remark 4.1 where this assumption is relaxed). We do not use the assumption that Y_+ is normal in this section. X will be assumed to be an order convex subset of Y and Φ is assumed to be a monotone semiflow on X. Additional assumptions are listed below for convenient reference:

(M) There exists $\tau > 0$ such that $\Phi_\tau(x_1) \ll \Phi_\tau(x_2)$ whenever $x_1, x_2 \in X$ and $x_1 < x_2$.
(D) $\Phi_\tau : X \to X$ is continuously differentiable on X. We write $\Phi'_\tau(x)$ for the derivative at x.
(S) $\Phi'_\tau(e)$ is compact (completely continuous) and $\Phi'_\tau(e)(Y_+ \setminus 0) \subset \text{Int} Y_+$ for each $e \in E$.

(M) is somewhat weaker than strong monotonicity but it implies the SOP property, (D) is a smoothness assumption, and (S) has implications for the spectrum of $\Phi'_\tau(e)$ by the well-known Krein-Rutman Theorem (see below). We say that $\Phi'_\tau(e)$ is *strongly positive* if the second of assumptions (S) holds.

Observe that if $x \in X$, $y \in Y_+$, $h > 0$, and $x + hy \in X$ then (M) and (D) imply

$$\frac{\Phi_\tau(x+hy) - \Phi_\tau(x)}{h} \ge 0$$

so on taking the limit as $h \to 0$, we get $\Phi'_\tau(x)y \ge 0$. In words, $\Phi'_\tau(x)$ is a positive operator in the sense that $\Phi'_\tau(x)Y_+ \subset Y_+$. Therefore, the assumption that $\Phi'_\tau(e)$ is strongly positive is not such a severe one and, in fact, one usually must verify it anyway to prove that Φ_τ satisfies (M).

If (S) holds and $e \in E$, let $\sigma(\Phi'_\tau(e))$ be the spectrum of the linear operator $\Phi'_\tau(e)$. Since it is a compact operator, the spectrum consists of an at most countable set of eigenvalues plus possibly $\{0\}$ and the eigenvalues have no accumulation point except possibly zero. Let $\rho(e)$ be the spectral radius of $\Phi'_\tau(e)$, that is,

$$\rho(e) = \max\{|\lambda| : \lambda \in \sigma(\Phi'_\tau(e))\}.$$

A consequence of the well-known Krein-Rutman Theorem is that $\rho(e)$ is a simple eigenvalue of $\Phi'_\tau(e)$ and there is a corresponding positive eigenvector. The result is stated precisely below. Here, $N(A)$ denotes the null space of the linear operator A.

THEOREM 4.1. *(Krein-Rutman) Let A be a compact, strongly positive linear operator on the Banach space Y. Then $r = \rho(A)$, the spectral radius of A, is a positive eigenvalue of A and*

$$N(A - rI) = \cup_{k \geq 1} N((A - rI)^k) = span\{z\} \tag{4.1}$$

where $z \gg 0$. If $v \in Y_+$ is any other eigenvector of A, then v is a positive multiple of z. Finally, $|\lambda| < r$ for all $\lambda \in \sigma(A)$, $\lambda \neq r$.

The eigenvalue r is said to have *algebraic multiplicity one* if (4.1) holds. See Krein and Rutman(1950) or Zeidler(1986) for a proof of the Krein-Rutman Theorem. Theorem 4.1 leads to the following result.

LEMMA 4.2. *For each $e \in E$, $\rho(e) \in \sigma(\Phi'_\tau(e))$ and there exists a unique unit norm eigenvector $z(e) \gg 0$ such that*

$$\Phi'_\tau(e) z(e) = \rho(e) z(e).$$

Moreover, the maps $e \to z(e)$ and $e \to \rho(e)$ are continuous on E.

PROOF. The first assertions follow from the Krein-Rutman Theorem. The upper semicontinuity of the spectral radius follows from the upper semicontinuity of the spectrum as a function of the operator (Kato(1976)). The lower semicontinuity follows from the lower semicontinuity of isolated parts of the spectrum (Kato(1976), Thm.3.1, Remark 3.3, Thm. 3.16, Chap. IV). Let $P(e)$ be the projection onto $N(\Phi'_\tau(e) - \rho(e)I)$ along $Im(\Phi'_\tau(e) - \rho(e)I)$. From Kato(1976) (Thm. 3.16, Chap. IV), the map $e \to P(e)$ is continuous on E. Let $e_n \to e$ where $e, e_n \in E$ and let $z_n = z(e_n)$, $z = z(e)$. Then $(I - P(e))z_n = (P(e_n) - P(e))z_n \to 0$ as $n \to \infty$, and $\{P(e)z_n\}$ is precompact so $\{z_n\}$ is precompact. If $z_{n_i} \to u$ for some subsequence, then

$$\Phi'_\tau(e) u = \lim \Phi'_\tau(e_{n_i}) z_{n_i} = \lim \rho(e_{n_i}) z_{n_i} = \rho(e) u.$$

It follows from the uniqueness of the positive eigenvector for $\Phi'_\tau(e)$ (Theorem 4.1) that $u = z$ and that $z_n \to z$ as $n \to \infty$. $\square$

A principle goal will be to show that most quasiconvergent points are convergent. The next result says that every equilibrium in the limit set of a quasiconvergent point that is not a convergent point, is unstable in the linear approximation.

LEMMA 4.3. *Let $\omega(x) \subset E$ be a compact limit set. If $\omega(x)$ is not a singleton, then $\rho(e) > 1$ for all $e \in \omega(x)$. In particular, there exists $\gamma > 1$ such that $\rho(e) > \gamma$ for all $e \in \omega(x)$.*

PROOF. Suppose that $\omega(x)$ is not a singleton. Since $\omega(x)$ is connected, every point e of it is a point of accumulation of $\omega(x)$. Let $e_n \in \omega(x)$ satisfy $e_n \neq e$ and $e_n \to e$. Then

$$e - e_n = \Phi_\tau(e) - \Phi_\tau(e_n) = \Phi'_\tau(e)(e - e_n) + o(|e - e_n|)$$

where $o(|e - e_n|)/|e - e_n| \to 0$ as $n \to \infty$. Put $v_n = (e - e_n)/|e - e_n|$. Then

$$v_n = \Phi'_\tau(e) v_n + r_n, \quad r_n \to 0, \quad n \to \infty.$$

The compactness of $\Phi'_\tau(e)$ implies that v_n has a convergent subsequence v_{n_i} and passing to the limit along this subsequence leads to $v = \Phi'_\tau(e)v$ for some vector v with unit norm. Thus $\rho(e) \geq 1$. If $\rho(e) = 1$, then by the Theorem 4.1, $v = rz(e)$ where $r = \pm 1$. Consequently,

$$(e - e_{n_i})/|e - e_{n_i}| \to rz(e)$$

as $i \to \infty$. It follows that $e \ll e_{n_i}$ or $e \gg e_{n_i}$ for all large i, contradicting the Nonordering of Limit Sets. We conclude that $\rho(e) > 1$ for every $e \in \omega(x)$. The final assertion follows from the compactness of $\omega(x)$ and the continuity of spectral radius (Lemma 4.2). □

It will be convenient to consider another norm on Y. If $w \in \text{Int} Y_+$ is fixed, then $U = \{y \in Y : -w \ll y \ll w\}$ is an open neighborhood of the origin. Consequently, if $y \in Y$, then there exists $t_0 > 0$ such that $t_0^{-1} y \in U$, or, $-t_0 w \ll y \ll t_0 w$. Define the w-norm by

$$\|y\|_w = \inf\{t > 0 : -tw \leq y \leq tw\}.$$

Since $w \in \text{Int} Y_+$, there exists $\delta > 0$ such that for all $y \in Y$, $y \neq 0$, it follows that $w \pm \delta \frac{y}{|y|} \in Y_+$. Thus,

$$\|y\|_w \leq \delta^{-1} |y|$$

holds for all $y \in Y$, implying that the w-norm is weaker than the original norm. In fact, the two norms are equivalent if Y_+ is normal but we will have no need for this result. See Amann(1976) for more results in this direction.

It will be useful to renormalize the positive eigenvector $z(e)$ of $\Phi'_\tau(e)$ for $e \in E$. The next result says this can be done continuously. Continuity is always to be understood, unless explicitly stated otherwise, with respect to the original norm topology.

LEMMA 4.4. *Let $Z(e) = z(e)/\|z(e)\|_w$ and $\beta(e) = \sup\{\beta > 0 : Z(e) \geq \beta w\}$. Then $\beta(e) > 0$, $Z(e) \geq \beta(e)w$ and the maps $e \to Z(e)$ and $e \to \beta(e)$ are continuous.*

PROOF. Since the w-norm is weaker than the original norm, the map $e \to \|z(e)\|_w$ is continuous on E. This implies that $Z(e)$ is continuous on E. It is easy to see that $\beta(e) > 0$. Let $\epsilon > 0$ satisfy $2\epsilon < \beta(e)$ and let $e_n \in E$ satisfy $e_n \to e$ as $n \to \infty$. Then $-\epsilon w \leq Z(e) - Z(e_n) \leq \epsilon w$ for all large n by continuity of Z and because the w-norm is weaker than the original norm. Therefore, $Z(e_n) = Z(e_n) - Z(e) + Z(e) \geq (\beta(e) - \epsilon)w$, so $\beta(e_n) \geq \beta(e) - \epsilon$ for all large n. Similarly, $Z(e) = Z(e) - Z(e_n) + Z(e_n) \geq (\beta(e_n) - \epsilon)w$ for all large n, so $\beta(e) \geq \beta(e_n) - \epsilon$ for all large n. Thus, $\beta(e) - \epsilon \leq \beta(e_n) \leq \beta(e) + \epsilon$ holds for all large n, completing the proof. □

The key to improving Theorem 1.4.3 is to sharpen the Limit Set Dichotomy. The additional assumptions (M),(D), and (S) allow us to do that.

THEOREM 4.5. *Assume that (M),(D) and (S) hold and that $x_1, x_2 \in X$ satisfy $x_1 < x_2$ and $O(x_1), O(x_2)$ have compact closure in X. Then either*
(a) $\omega(x_1) < \omega(x_2)$, *or*
(b) $\omega(x_1) = \omega(x_2) = \{e\}$ *for some $e \in E$.*

PROOF. By the Limit Set Dichotomy, it suffices to show that if $\omega(x_1) = \omega(x_2)$, then (b) holds. $\omega(x_1)$ is compact, connected and $z_1 < z_2$ cannot hold for any pair of points $z_1, z_2 \in \omega(x_1)$ by the Nonordering of Limit Sets. Let $v_n = \Phi_{n\tau}(x_1)$, $u_n = \Phi_{n\tau}(x_2)$, $S(x) \equiv \Phi_\tau(x)$ and $K = \omega(x_1)$. Then K is a set of fixed points of S, $v_{n+1} = Sv_n$, $u_{n+1} = Su_n$, $u_n \gg v_n$ for $n \geq 1$, and $\text{dist}(K, u_n)$, $\text{dist}(K, v_n) \to 0$ as $n \to \infty$. Suppose that K contains more than a single element.

Fix $w \in \text{Int} Y_+$ and define $\alpha_n = \sup\{\alpha > 0 : u_n \geq v_n + \alpha w\}$. Then $\alpha_n > 0$ and $u_n \geq v_n + \alpha_n w$. We show that $\alpha_n \to 0$ as $n \to \infty$. Otherwise, there is a subsequence α_{n_i} satisfying $\alpha_{n_i} \geq \alpha > 0$. By the compactness of K and the fact that $u_n, v_n \to K$, we may as well assume that $u_{n_i} \to u$ and $v_{n_i} \to v$ where $u, v \in K$. Then $u \geq v + \alpha w$ so $u > v$, in contradiction to the Nonordering of Limit Sets. Therefore, $\alpha_n \to 0$.

Choose $e_n \in K$ such that $v_n - e_n \to 0$ as $n \to \infty$. Since K is not a singleton, then by Lemma 4.3, there exists $r > 1$ such that $\rho(e) > r$ for all $e \in K$. Let $z_n = Z(e_n)$ be the normalized positive eigenvector for $S'(e_n) = \Phi'_\tau(e_n)$ so $\|z_n\|_w = 1$ and $z_n \leq w$. By Lemma 4.4, $\beta(e)$ is continuous in $e \in E$ and since K is a compact subset of E, there exists $\epsilon > 0$ such that $\beta(e) \geq \epsilon$ for all $e \in K$. In particular, $w \geq z_n \geq \epsilon w$ for all n.

Choose l, a positive integer, such that $r^l \epsilon > 1$. Then, since $u_n \geq v_n + \alpha_n w \gg v_n$ and X is order convex, $v_n + \eta \alpha_n w \in X$ for $0 \leq \eta \leq 1$ and the fundamental theorem of calculus leads to

$$S^l(v_n + \alpha_n w) - S^l v_n = (S^l)'(e_n) \alpha_n w + \alpha_n \delta_n$$

where S^l denotes the composition of S with itself l-times and

$$\delta_n = \int_0^1 [(S^l)'(v_n + \eta \alpha_n w) - (S^l)'(e_n)] w \, d\eta.$$

Using that $v_n + \alpha_n w - e_n \to 0$, K is compact, and $(S^l)'$ is continuous, it is easy to show that

$$\max_{0 \leq \eta \leq 1} |[(S^l)'(v_n + \eta \alpha_n w) - (S^l)'(e_n)] w| \to 0, \quad n \to \infty.$$

It follows that $d_n = \|\delta_n\|_w \to 0$ as $n \to \infty$. Using $w \geq z_n \geq \epsilon w$ and $\delta_n \geq -d_n w$, we have

$$\begin{aligned} S^l(v_n + \alpha_n w) - S^l v_n &\geq [(S^l)'(e_n)] \alpha_n w - \alpha_n d_n w \\ &\geq [S'(e_n)]^l \alpha_n w - \alpha_n d_n w \\ &\geq r^l \alpha_n z_n - \alpha_n d_n w \\ &\geq (r^l \epsilon - d_n) \alpha_n w \\ &\geq \alpha_n w \end{aligned}$$

for all large n. Therefore,

$$u_{n+l} = S^l(u_n) \geq S^l(v_n + \alpha_n w) \geq S^l v_n + \alpha_n w = v_{n+l} + \alpha_n w$$

for all large n. This implies that

$$\alpha_{n+l} \geq \alpha_n$$

for all large n, contradicting that $\alpha_n \to 0$ as $n \to \infty$. This contradiction implies that K is a singleton. $\square$

The strengthened Limit Set Dichotomy implies that if x_1 is a quasiconvergent point that is not convergent and $x_1 < x_2$, then (a) holds. Theorem 4.5 makes possible a strengthening of the Sequential Limit Set Trichotomy. Here, we need the compactness hypothesis (C) of Chapter 1.

PROPOSITION 4.6. *Assume* (M),(D),(S) *and* (C) *hold. Let $x_0 \in X$ be approximated from below in X by the sequence $\tilde{x}_n$. Then there exists a subsequence x_n of $\tilde{x}_n$ such that $x_n < x_{n+1} < x_0$, $n \geq 1$, $x_n \to x_0$, and $\{x_n\}$ satisfies one of the alternatives* (a) *or* (b) *of Theorem 1.4.1, or*
(c) *There exists $u_0 \in E$ such that $\omega(x_n) = \omega(x_0) = u_0$ for $n \geq 1$.*

PROOF. This follows immediately, using Theorem 4.5 in the proof of Theorem 1.4.1. □

The main result of this section can now be stated. It says that a generic point of X is a convergent point.

THEOREM 4.7. *Suppose that* (M),(D),(S), *and* (C) *hold and suppose that each point of X can be approximated either from above or from below in X. Then $\mathrm{Int}C$ is dense in X.*

PROOF. The proof is identical to that of Theorem 1.4.3 except that it is supposed that $x_0 \in X \setminus \mathrm{Int}C$. The sequence y_n in that proof belongs to $X \setminus C$. The proof of Theorem 1.4.3 then leads to the conclusion that $x_0 \in \overline{\mathrm{Int}C}$. □

A serious drawback of Theorem 4.7 is the assumption that the cone Y_+ has nonempty interior in Y. This assumption severely limits the application of the result to infinite dimensional systems. In the remark to follow, we describe a generalization of Theorem 4.7 which does not require this assumption. It is in the spirit of Remark 1.1 of the previous chapter.

REMARK 4.1. Let Y be a Banach space with cone Y_+ having nonempty interior in Y but we do not assume that $X \subset Y$. Instead, X is an ordered metric space with metric d as in Chapter 1 such that $Z = X \cap Y$ is nonempty, order convex as a subset of Y, and such that the identity map $z \in (Z, d) \to z \in (Z, d_Y)$ is continuous from Z, endowed with the metric of X, into Z, endowed with the metric induced by the norm on Y. In other words, the latter metric is stronger than d. Furthermore, we assume that the restriction of the order relation $\leq_Y$ on Y to Z agrees with the restriction of the order relation $\leq$ on X to Z. As usual, it is assumed that Φ is a semiflow on X, each point of X can be approximated from above or from below in X and that (C) holds. Hypotheses (M),(D) and (S) above are replaced by the following:
(I) $\Phi_t(Z) \subset Z$ for $t \geq 0$.
(J) There exists $\tau > 0$ such that $\Phi_\tau(X) \subset Z$ and $\Phi_\tau : X \to Z$ is continuous.
(M) If $x_1, x_2 \in X$ satisfy $x_1 < x_2$, then $\Phi_\tau(x_1) \ll_Y \Phi_\tau(x_2)$ where τ is as in (J) and the strong order $\ll_Y$ is that induced by $\mathrm{Int}Y_+$ on Z.
(D) For any $e \in E$, there exists a neighborhood U of e in Z, in the topology inherited from Y, such that Φ_τ is continuously differentiable on U, as a map from U into Z with respect to the norm topology of Y.
(S) For each $e \in E$, the Frechet derivative $\Phi'_\tau(e)$, which exists by (D), is compact and strongly positive on Y_+.

Then, with these hypotheses, the conclusion of Theorem 4.7 holds.

5. Unstable Equilibria and Connecting Orbits

The purpose of this section is to provide an abstract treatment of some ideas which are quite useful in the applications. The reader is advised to skip over this section on a first reading. Beginning in Chapter 4, the text will refer back to ideas developed here and the reader may read it then. Basically, we are concerned with an unstable equilibrium and we will show that typically one can find a stable equilibrium greater (smaller) than the unstable one by following a monotone full orbit of the semiflow issuing from the unstable equilibrium.

We begin by describing the basic setup and introducing some required definitions. Let Y be a Banach space with a cone Y_+ which has nonempty interior, $\text{Int} Y_+$, in Y. Occasional use is made of the simple observation that if $0 \ll y$ ($y \in \text{Int} Y_+$) and $z \geq y$, then $0 \ll z$. Let ∂Y_+ denote the boundary of Y_+. We suppose that X is a nonempty subset of Y and that Φ is a *strictly monotone semiflow* on X. By this it is meant that $\Phi_t(x) < \Phi_t(y)$ whenever x and y belong to X and $x < y$. Obviously, a strictly monotone semiflow is also a monotone semiflow. We assume throughout this section that the orbit of each point of X has compact closure in X.

A point $x \in X$ is a *sub-equilibrium* if $\Phi_t(x) \geq x$ for all $t \geq 0$ and a *strict sub-equilibrium* if $\Phi_t(x) > x$ for all $t > 0$. A *super-equilibrium* and *strict super-equilibrium* are defined similarly with the inequalities reversed. Let E_+ be the set of all sub-equilibria, E_{++} be the set of all strict sub-equilibria, E_- be the set of super-equilibria and E_{--} be the set of all strict super-equilibria. Obviously,

$$E = E_+ \cap E_-.$$

The next result summarizes some of the properties of these sets.

LEMMA 5.1. *E_+ is closed and positively invariant and $E_{++} = E_+ \setminus E$. If $x \in E_{++}$ and $0 \leq t < s$, then $\Phi_t(x) < \Phi_s(x)$ and there exists $e \in E$ such that*

$$\Phi_t(x) \to e, \quad t \to \infty.$$

Analogous assertions hold for E_- and E_{--} with inequalities reversed.

PROOF. E_+ is closed because Y_+ is closed and Φ is continuous. Positive invariance of E_+ follows from monotonicity of Φ. Obviously, $E_{++} \subset E_+ \setminus E$. If $x \in E_+ \setminus E$, then $\Phi_t(x) \geq x$ for all $t \geq 0$ and the equality $\Phi_t(x) = x$ cannot hold for all $t > 0$. In fact, $\Phi_t(x) > x$ for all small $t > 0$. For if equality holds for $t = t_n > 0$ where $t_n \to 0$ as $n \to \infty$ and $t > 0$, then we may write $t = q_n t_n + r_n$ where q_n is a nonnegative integer and $0 \leq r_n < t_n$. Therefore, $\Phi_t(x) = \Phi_{r_n}(\Phi_{q_n t_n}(x)) = \Phi_{r_n}(x) \to x$ as $n \to \infty$, so $x \in E$. Thus, either $\Phi_t(x) > x$ holds for all $t > 0$, as asserted, or there exists $t_0 > 0$ such that $\Phi_t(x) > x$ for $0 < t < t_0$ and $\Phi_{t_0}(x) = x$. But in this case, $\Phi_{t_0/2}(x) > x$ and since Φ is strictly monotone we have that $\Phi_{t_0}(x) > \Phi_{t_0/2}(x) > x$, a contradiction. We have proved that $E_+ \setminus E \subset E_{++}$.

If $x \in E_{++}$, then $\Phi_t(x) > x$ for all $t > 0$ and strict monotonicity implies that $\Phi_{t+s}(x) > \Phi_s(x)$ for all $t, s > 0$. Therefore, $\Phi_t(x) < \Phi_s(x)$ whenever $0 \leq t < s$. As $O(x)$ has compact closure in X, it follows from Lemma 1.1.2 that the sequence $\Phi_{t_n}(x)$ converges to a point of $\omega(x)$ for each increasing sequence t_n such that $t_n \to \infty$. If t_n and s_n are any two unbounded increasing sequences, then there is a increasing and unbounded sequence r_n having t_n and s_n as subsequences. As $\Phi_{r_n}(x)$ converges to a point of $\omega(x)$ as $n \to \infty$, it follows that the sequences $\Phi_{t_n}(x)$ and $\Phi_{s_n}(x)$ converge to the same limits. Thus $\omega(x)$ is a single point. □

5. UNSTABLE EQUILIBRIA AND CONNECTING ORBITS

For the remainder of this section we fix $x_0 \in E$. As we will see repeatedly in later chapters, if x_0 is unstable in a suitably strong sense, then we can find strict sub-equilibria, x, arbitrarily close to x_0 satisfying $x_0 \ll x$. Similarly, we can find strict super-equilibria, x, arbitrarily close to x satisfying $x \ll x_0$. We will consider the former case in detail. The latter case is obtained from the former by reversing inequalities. Our main assumption is that there exists a monotone arc of sub-equilibria through x_0 in the following sense:

(U_+) There exists a continuous map $\eta : [0,1] \to E_+$ satisfying $\eta(0) = x_0$, $\eta(s) \in E_{++}$ for $s > 0$, and $x_0 = \eta(0) \ll \eta(s_1) \ll \eta(s_2)$ whenever $0 < s_1 < s_2 \le 1$. Set $\Gamma = \eta([0,1])$.

In the applications, Γ is a line segment through x_0 in the direction of the principal eigenvector of the generator of the linearized semiflow about x_0. The existence of such an eigenvector usually follows from the Krein-Rutman theorem. The next result says that if (U_+) holds, then there is a unique equilibrium greater than x_0 to which the solutions through each point of Γ approach monotonically. Furthermore, this equilibrium has a substantial domain of attraction.

PROPOSITION 5.2. *Let (U_+) hold and suppose that Φ is strongly monotone on $\{x \in X : x_0 \ll x\}$. Then there exists an equilibrium e satisfying $x_0 \ll e$ with the property that $\Phi_t(x) \to e$ as $t \to \infty$ for all $x \in \Gamma \setminus \{x_0\}$. Furthermore, if $x \in \Gamma$, $x \ne x_0$ and $0 < t_1 < t_2$, then $x \ll \Phi_{t_1}(x) \ll \Phi_{t_2}(x)$. Moreover, if $x \in X$ satisfies $x_0 \ll x < e$, then $\Phi_t(x) \to e$ as $t \to \infty$.*

PROOF. Fix $y \in \Gamma$ with $y \ne x_0$. Since $y \in E_{++}$, it follows from Lemma 5.1 that $\Phi_{t_1}(y) < \Phi_{t_2}(y)$ whenever $0 \le t_1 < t_2$ and furthermore, $\Phi_t(y)$ converges to an equilibrium e_y as $t \to \infty$. For $t > 0$, $\Phi_{t/2}(y) > y \gg x_0$ and by the strong monotonicity assumption it follows that $\Phi_t(y) \gg \Phi_{t/2}(y) > y$ so $\Phi_t(y) \gg y$. Applying Φ_s to the previous inequality results in $\Phi_{t+s}(y) \gg \Phi_s(y) \gg y$ for $s, t > 0$

As $y \in \Gamma$, $y \ne x_0$ was arbitrary, we find that for each such y there is an equilibrium e_y such that $\Phi_t(y) \to e_y$ monotonically as $t \to \infty$. We now show that there is an equilibrium e such that $e = e_y$ for all such y. Since $y_1 = \eta(s_1) \ll \eta(s_2) = y_2$ if $s_1 < s_2$, the strict monotonicity of Φ implies that $\Phi_t(y_1) < \Phi_t(y_2)$ so on letting $t \to \infty$ we conclude that $e_{y_1} \le e_{y_2}$. Again, fix $s \in (0,1]$ and set $y = \eta(s)$. Because $\eta(s) \ll \Phi_{t_0}(\eta(s)) \ll \Phi_{t+t_0}(\eta(s))$ for $t, t_0 > 0$, it follows that $\eta(s) \ll e_y$ by letting $t \to \infty$ in the previous inequality. Let $Q = \{h \ge 0 : s + h \le 1$ and $\eta(s+h) \le e_y\}$. Since $\eta(s) \ll e_y$ and η is continuous and monotone, Q is nonempty, closed and has the property that if $h > 0$ belongs to Q, then $[0,h] \subset Q$. Thus $Q = [0, h^*]$ for some $h^* > 0$. Consequently, $\eta(s + h^*) \le e_y$ and $\Phi_t(\eta(s+h^*)) \le \Phi_t(e_y)$. Letting $t \to \infty$, we conclude that $e_{\eta(s+h^*)} \le e_y$. But $e_{\eta(s+h^*)} \ge e_y$ was noted above ($\eta(s+h^*) \gg \eta(s) = y$) so therefore, $e_y = e_{\eta(s+h^*)}$. As a consequence, $e_y = e_{\eta(s+h^*)} \gg \eta(s+h^*)$. If $s + h^* < 1$, then by the continuity of η we could find $k > h^*$ such that $s + k < 1$ and $\eta(s+k) \le e_y$ so $k \in Q$. This contradicts that $Q = [0, h^*]$ and therefore, we conclude that $s + h^* = 1$. Thus $e_{\eta(s)} = e_{\eta(1)}$ and as $s \in (0,1]$ was arbitrary, this holds for every such s. Let $e = e_{\eta(1)}$.

Finally, if $x \in X$ satisfies $x_0 \ll x < e$, then $x_0 \ll \eta(s) \ll x$ for all small positive s. Consequently, for such s, $\Phi_t(\eta(s)) < \Phi_t(x) < e$ for $t > 0$ by strict monotonicity. Since $\Phi_t(\eta(s)) \to e$ as $t \to \infty$, the same must hold for $\Phi_t(x)$. □

An analogous result holds if we replace the assumption (U_+) by the following assumption.

(U_-) There exists a continuous map $\eta : [0,1] \to E_-$ satisfying $\eta(0) = x_0$, $\eta(s) \in E_{--}$ for $s > 0$, and $\eta(s_2) \ll \eta(s_1) \ll \eta(0) = x_0$ whenever $0 < s_1 < s_2 \leq 1$. Set $\Gamma = \eta([0,1])$.

In this case we would assume that Φ is strongly monotone on $\{x \in X : x \ll x_0\}$. We would then conclude the existence of an equilibrium e satisfying $e \ll x_0$ with the property that $\Phi_t(x) \to e$ as $t \to \infty$ for all $x \in \Gamma$ except $x = x_0$. For such an x and $0 < t_1 < t_2$ it follows that $\Phi_{t_2}(x) \ll \Phi_{t_1}(x) \ll x$. Furthermore, e attracts all solutions starting at x where $e < x \ll x_0$.

The assumption that Φ is strongly monotone on the set $\{x \in X : x_0 \ll x\}$ in Proposition 5.2 can be replaced by the assumption that there exists an equilibrium $\hat{e}$ satisfying $x_0 \ll \hat{e}$ and that Φ is strongly monotone on the set $\{x \in X : x_0 \ll x \ll \hat{e}\}$. In this case, it is easily shown that $e \leq \hat{e}$ where equality $e = \hat{e}$ is possible. In the case that (U_-) is assumed, then the assumption that Φ is strongly monotone on $\{x \in X : x \ll x_0\}$ can be replaced by the assumption that there exists an equilibrium $\hat{e}$ satisfying $\hat{e} \ll x_0$ and such that Φ is strongly monotone on the set $\{x \in X : \hat{e} \ll x \ll x_0\}$.

Let $A \subset X$ and $x \in A$. A continuous function $\phi : \mathbb{R} \to A$ is called a *full orbit of Φ in A through x* if $\phi(0) = x$ and $\Phi_t(\phi(s)) = \phi(t+s)$ for all $t \geq 0$ and $s \in \mathbb{R}$. Thus $\phi(t) = \Phi_t(x)$ for $t \geq 0$ and for $t < 0$, $\phi(t)$ is a backward solution through x in A. The reader is cautioned that there may be more than one backward solution through x in A if Φ is not one-to-one. It is well-known that if A is an invariant set for Φ, then through each point x of A there is at least one full orbit of Φ in A through x. In fact, there is a point $x_{-1} \in A$ such that $\Phi_1(x_{-1}) = x$, a point $x_{-2} \in A$ such that $\Phi_1(x_{-2}) = x_{-1}$. Continuing in this manner, for every natural number n, there is a point $x_{-n} \in A$ such that $\Phi_1(x_{-n}) = x_{-n+1}$, where $x_0 = x$. For $-n \leq t < -n+1$ define $\phi(t) = \Phi_{t+n}(x_{-n})$ and $\phi(t) = \Phi_t(x)$ for $t \geq 0$. It is easy to see that ϕ is a full orbit of Φ in A through x. See Ladyzhenskaya(1991) for further details. A typical way of finding a compact invariant set is as follows. Suppose that Γ is a nonempty subset of X and let $O(\Gamma) = \cup_{x \in \Gamma} O(x)$ be the orbit of Γ. The omega limit set of Γ is defined by

$$\omega(\Gamma) = \cap_{s>0} \overline{\cup_{t \geq s} \Phi_t(\Gamma)}.$$

If, for example, $O(\Gamma)$ is bounded and Φ_{t_0} is completely continuous, then (see Conley(1978), Ladyzhenskaya(1991), or Hale(1988)) $\omega(\Gamma)$ is nonempty, compact, invariant, connected if Γ is connected, and $\Phi_t(\Gamma)$ tends to $\omega(\Gamma)$ as $t \to \infty$. Clearly, $y \in \omega(\Gamma)$ if and only if there exists sequences $\{x_n\} \subset \Gamma$ and $\{t_n\}$ satisfying $t_n \to \infty$ such that $\Phi_{t_n}(x_n) \to y$ as $n \to \infty$.

The next result is the main one of this section. Building on Proposition 5.2, we show the existence of a monotone full orbit connecting x_0 to e. Our hypotheses will appear quite strange on first reading. To understand them, keep in mind that in the applications we quite often have very good knowledge of the behavior of solutions starting at points $x \in X$ belonging to $x_0 + \partial Y_+$. Therefore, we are free to exploit our knowledge of the behavior of solutions beginning on this set by making hypotheses about the existence or non-existence of certain kinds of solutions beginning on this set.

5. UNSTABLE EQUILIBRIA AND CONNECTING ORBITS

THEOREM 5.3. *In addition to the hypotheses of Proposition 5.2, assume that each of the following hold:*
(1) $O(\Phi_{t_0}(\Gamma))$ *has compact closure in* X *for some* $t_0 > 0$.
(2) *There exists a neighborhood* V *of* x_0 *such that no equilibrium in* V *other than* x_0 *belongs to* $x_0 + \partial Y_+$.
(3) *If* $x \in (x_0 + \partial Y_+) \cap X$, *then either* (a) $\Phi_t(x) \in x_0 + \partial Y_+$ *for* $t \geq 0$, *or* (b) $x_0 \ll \Phi_t(x)$ *for* $t > 0$.
(4) *There does not exist a full orbit* ϕ *in* $(x_0 + \partial Y_+) \cap E_{++}$ *satisfying* $\phi(t) \to x_0$ *as* $t \to -\infty$.

Then there exists a full orbit ϕ *in* $E_{++} \cap \omega(\Gamma)$ *satisfying*

$$\phi(t) \to x_0, \quad t \to -\infty$$

and

$$\phi(t) \to e, \quad t \to +\infty$$

where e *is the equilibrium of Proposition 5.2. Furthermore, whenever* $s_1, s_2 \in \mathbb{R}$ *satisfy* $s_1 < s_2$, *then*

$$x_0 \ll \phi(s_1) \ll \phi(s_2) \ll e.$$

PROOF. The key ideas of the proof are as follows. Since $\Gamma \subset E_+$, $J \equiv \omega(\Gamma)$ satisfies $J \subset E_+$ because E_+ is closed and positively invariant. Furthermore, J is nonempty, compact, invariant and connected (since Γ is connected). Clearly, x_0 and e belong to J. If $x \in J \setminus E$, then $x \in E_{++}$ by Lemma 5.1. As noted above, there is a full orbit ϕ of Φ in J through x. Since $x \in E_{++}$ and $J \subset E_+$, it follows that the image of ϕ belongs to E_{++}. Therefore, if $s_1, s_2 \in \mathbb{R}$ and $s_1 < s_2$, then $\phi(s_2) = \Phi_{s_2-s_1}(\phi(s_1)) > \phi(s_1)$. Consequently, $\phi(s)$ is increasing and, as its image belongs to the compact set J, the limits $\phi(+\infty) = \lim_{t \to \infty} \phi(t)$ and $\phi(-\infty) = \lim_{t \to -\infty} \phi(t)$ exist and belong to $E \cap J$ by arguments similar to those used in the proof of Lemma 5.1. Indeed, for $t \geq 0$, $\Phi_t(\phi(-\infty)) = \lim_{s \to -\infty} \Phi_t(\phi(s)) = \lim_{s \to -\infty} \phi(s+t) = \phi(-\infty)$, so $\phi(-\infty) \in E$. Therefore, every nonequilibrium point of J lies on an increasing full orbit of Φ in J which connects two equilibria.

If $x \in J$ then there exist sequences $s_n \in [0,1]$ and $t_n \to \infty$ such that $\Phi_{t_n}(x_n) \to x$ where $x_n = \eta(s_n)$. Since $x_0 \leq x_n \ll e$ it follows from monotonicity of Φ that $x_0 \leq x \leq e$, or $x_0 \leq J \leq e$. If $e_0 \in E \cap J$ and $x_0 \ll e_0$, then $x_0 \ll \eta(s) \ll e_0$ for all small s. Therefore, $x_0 \ll \Phi_t(\eta(s)) < e_0$ by monotonicity of Φ so on letting $t \to \infty$ we conclude from Proposition 5.2 that $e \leq e_0$. But $e_0 \leq e$ was noted above, hence $e_0 = e$. We conclude that e is the unique equilibrium belonging to J that satisfies $x_0 \ll e$.

We note that the neighborhood V in assumption (2) can be chosen to be given by $V = \{y \in Y : y_1 \ll y \ll \eta(s_0)\}$ for some $y_1 \in Y$ satisfying $y_1 \ll x_0$ and for some $s_0 \in (0,1]$ since such a neighborhood is contained in any neighborhood of x_0. Hereafter, we assume that V is chosen as above. Then $z \in V$ whenever there exists $y \in V$ such that $x_0 \leq z \leq y$.

Let $r > 0$ satisfy $r < |e - x_0|$ be such that the closed ball B of radius r about x_0 in Y is contained in the neighborhood V of assumption (2). As J is connected and the interior of B contains x_0 but not e, there exists $x \in J$ belonging to the boundary of B ($|x - x_0| = r$). Of course, $x_0 < x < e$ and $x \notin E$ since either $x \in x_0 + \partial Y_+$ or $x_0 \ll x$ and in the first case, $x \notin E$ by assumption (2), and in the second case $x \notin E$ since $x \neq e$. Thus $x \in E_{++}$ and there is an increasing full orbit ϕ in J through x with the property that $\phi(-\infty)$ and $\phi(+\infty)$ are equilibria in

J. We wish to show that $\phi(-\infty) = x_0$ and $\phi(+\infty) = e$. Now $\phi(-\infty) < x < e$ so the equilibrium $\phi(-\infty) \ne e$ and therefore $\phi(-\infty)$ must belong to $x_0 + \partial Y_+$. Since $x_0 \le \phi(-\infty) < x$ and $x \in V$, it follows that $\phi(-\infty) \in V$. But then $\phi(-\infty) = x_0$ by assumption (2). Now, if $x_0 \ll \phi(s)$ for some s, then $x_0 \ll \phi(t)$ for all $t \in \mathbb{R}$ by (3) and the fact that ϕ is an increasing full orbit. Thus either $\phi(t) \in x_0 + \partial Y_+$ for all $t \in \mathbb{R}$ or $x_0 \ll \phi(t)$ for all $t \in \mathbb{R}$. The former case contradicts assumption (4) so we conclude that $x_0 \ll \phi(t)$ for all $t \in \mathbb{R}$. But then it follows that $\phi(+\infty) = e$ since $x_0 \ll \eta(s) \ll x$ for all small $s > 0$ and $\Phi_t(\eta(s)) \to e$ as $t \to \infty$.

Finally, since $x_0 \ll \phi(s)$ for all $s \in \mathbb{R}$ and since if $s_1, s_2 \in \mathbb{R}$ satisfy $s_1 < s_2$ then $x_0 \ll \phi(s_1 - 1) < \phi(s_2 - 1)$, the strong monotonicity assumption implies that $\phi(s_1) = \Phi_1(\phi(s_1 - 1)) \ll \Phi_1(\phi(s_2 - 1)) = \phi(s_2)$. □

Hypothesis (1) can be weakened. All we really need is for $\omega(\Gamma)$ to be nonempty, compact, connected and to attract Γ (see Hale(1988)). Hypothesis (1) is satisfied if, for example, $O(\Gamma)$ is contained in a closed and bounded subset of X and Φ_{t_0} is completely continuous for some $t_0 > 0$.

A few remarks on the assumptions (2)-(4) will be helpful. They all concern the behavior of solutions starting at points of the set $(x_0 + \partial Y_+) \cap X$. Note that we allow the possibility that this set consists only of the single point x_0 in which case these hypotheses are trivially satisfied. Assumption (2) states that x_0 is the only equilibrium near x_0 belonging to this set. According to (3), the solution $\Phi_t(x)$ starting at a point of this set either remains in the set for all $t \ge 0$ or immediately leaves this set. Finally, (4) says that the set contains no increasing full orbit connecting x_0 to another equilibrium of this set. Assumptions (2)-(4) are automatically satisfied if Φ is strongly monotone on all of X.

An analogous result holds if instead of (U_+) we assume that (U_-) holds. In this case, hypotheses (2)-(4) concern solutions starting at points of $(x_0 - \partial Y_+) \cap X$. The reader should have no difficulty supplying these conditions.

6. Remarks and Discussion

The stability definitions are due to Hirsch(1988b), who uses the order topology generated by the w-norm instead of the original norm topology in his definitions. Proposition 1.1 and Theorem 1.2, taken from Smith and Thieme(1990a), are comparable to results of Hirsch(1988b) (see in particular, Proposition 9.5). The reader will find many more stability results in Hirsch(1988b) and in Smith and Thieme(1990a). While of theoretical importance, these results have not been used to a significant degree in the applications.

Proposition 2.1 asserts the existence of an equilibrium between two order-related asymptotically stable equilibria. One might expect that, generically, there is an unstable equilibrium between them, at least if the semiflow is continuously differentiable. This is proved in Smith(1986d) for ordinary differential equations. It is proved in greater generality for infinite dimensional, strictly order preserving maps in Dancer and Hess(1991) and Hess(1991). This result carries over to strictly order preserving semiflows as well. See Hess(1991).

A result similar to the Order Interval Trichotomy (Theorem 2.2) seems to have been first proved in Matano(1984) (Theorem 8). Instead of our alternative (ii) ((iii)), Matano's result asserts the existence of a monotone full orbit tending to u (v) as $t \to \infty$ and to v (u) as $t \to -\infty$. Hess(1991) obtains Matano's trichotomy assuming only that Φ is strictly monotone. Theorem 2.2 is essentially Corollary 3.12 in Smith and Thieme(1990a) and is similar to Hirsch(1988b) (Theorem 10.5).

Theorem 3.1, proved in Smith and Thieme(1990a), was patterned after Theorem 10.3 in Hirsch(1988b). Theorem 3.2 is also proved in Smith and Thieme(1990a).

Theorem 4.7 and the generalization contained in Remark 4.1 are special cases of Theorem 2.4 in Smith and Thieme(1991a). The approach in this paper was inspired by Poláčik(1989a) where the first generic convergence result was proved for abstract semilinear parabolic evolution systems. The setting of our result is more general and yields stronger conclusions when applied to the case treated in Poláčik(1989a), assuming less smoothness but more compactness.

Theorem 5.3 is an abstract version of a result of Hsu et al(1994) which in turn uses ideas contained in Matano(1984), Poláčik(1990), Dancer and Hess(1991) and Hess(1991). See also Smith (1986b). Some of Matano's work is presented in Leung(1989).

If most orbits with compact closure converge to an equilibrium, then we might next ask in what manner do they converge to equilibrium. It might be expected that most orbits converging to a stable equilibrium do so in an eventually monotone fashion. This has been shown in Mierszyński(1991) under quite general conditions for smooth strongly monotone dynamical systems, even if the equilibrium is not stable in the linear approximation. This work builds on earlier work of Poláčik(1989b).

CHAPTER 3

Competitive and Cooperative Differential Equations

0. Introduction

This chapter and the next one focus on ordinary differential equations in $\mathbb{R}^n$. A natural partial ordering on $\mathbb{R}^n$ is generated by the cone of vectors with nonnegative components but we shall see that the other orthants of $\mathbb{R}^n$ can equally serve to generate useful partial orderings. An ordinary differential equation can be solved both forward in time and backward in time. Roughly speaking, we will say that a system of ordinary differential equations is cooperative if it generates a monotone semiflow in the forward time direction and competitive if it generates a monotone semiflow in the backward time direction.

The chapter begins with the classical Kamke condition for an autonomous ordinary differential equation to generate a monotone semiflow in the forward time direction. Section two contains a useful result on positively invariant sets and monotone solutions. In addition, it is shown that all solutions of a planar competitive or cooperative system converge monotonically to equilibrium. The main results of the chapter are Theorem 3.4 and Theorem 4.1. The former asserts that the flow restricted to a compact limit set of a competitive or cooperative system in $\mathbb{R}^n$ is topologically equivalent to a flow on a compact invariant set of a Lipschitz system of differential equations in $\mathbb{R}^{n-1}$. In other words, the long term dynamics of an n-dimensional competitive or cooperative system can be no more badly behaved than that of a general system of one less dimension. Since our understanding of one and two dimensional (general) systems is fairly complete, we can expect to apply this result effectively to two and three dimensional competitive and cooperative systems. Theorem 4.1 establishes that the Poincaré-Bendixson Theorem holds for three dimensional competitive and cooperative systems.

Competitive systems in $\mathbb{R}^3$ can have attracting periodic orbits (cooperative systems cannot!) so additional attention is focused on these systems. It can be difficult to apply Theorem 4.1 even in the simple case where the system has only a single unstable equilibrium because of the potential existence of homoclinic orbits. In Theorem 4.2, sufficient conditions are given for all solutions that start off the (one-dimensional) stable manifold of the equilibrium to approach a nontrivial periodic orbit. Every periodic orbit of a three dimensional competitive or cooperative system contains an equilibrium "inside" it in a sense made precise in Proposition 4.3. These ideas are applied to the classical Field-Noyes model of the Belousov-Zhabotinsky reaction in section 6.

The strong order preserving property plays no role here. As a result, this chapter is essentially independent of the previous ones. In the following chapter, the irreducibility assumption on the Jacobian of the vector field is introduced in order to establish the strong order preserving property. This assumption is avoided in the present chapter.

1. The Kamke Condition

Consider the autonomous system of ordinary differential equations

$$x' = f(x) \tag{1.1}$$

where f is continuously differentiable on an open subset $D \subset \mathbb{R}^n$. We change our notation slightly to conform to more traditional notation for ordinary differential equations. For example, let $\phi_t(x)$ denote the solution of (1.1) that starts at the point x at $t = 0$. ϕ_t will be referred to as the *flow* corresponding to (1.1). We sometimes refer to f as the *vector field* generating the flow ϕ_t. If $\phi_t(x)$ is defined for all $t \geq 0$ ($t \leq 0$), then we write $\gamma^+(x) = \{\phi_t(x) : t \geq 0\}$ ($\gamma^-(x) = \{\phi_t(x) : t \leq 0\}$) for the *positive orbit* (*negative orbit*) through the point x. If $\gamma^-(x)$ has compact closure in D, then the *alpha limit set*, $\alpha(x)$, is defined in a similar way as the omega limit set:

$$\alpha(x) = \cap_{t \leq 0} \overline{\cup_{\tau \leq t} \phi_\tau(x)}.$$

The nonnegative cone in $\mathbb{R}^n$, denoted by $\mathbb{R}^n_+$, is the set of all n-tuples with nonnegative coordinates. It gives rise to a partial order on $\mathbb{R}^n$ by $y \leq x$ if $x - y \in \mathbb{R}^n_+$. Less formally, this is true if and only if $y_i \leq x_i$ for all i. We write $x < y$ if $x \leq y$ and $x_i < y_i$ for some i and we write $x \ll y$ if $x_i < y_i$ for all i. Our immediate goal is to provide sufficient conditions for ϕ to be a monotone dynamical system with respect to the ordering $\leq$. f is said to be of *type K* in D if for each i, $f_i(a) \leq f_i(b)$ for any two points a and b in D satisfying $a \leq b$ and $a_i = b_i$. The following classical result asserts that the type K condition is sufficient for the order preserving property to hold. It is also a necessary condition.

PROPOSITION 1.1. *Let f be type K on D and $x_0, y_0 \in D$. Let $<_r$ denote any one of the relations $\leq, <$ or $\ll$. If $x_0 <_r y_0$, $t > 0$ and $\phi_t(x_0)$ and $\phi_t(y_0)$ are defined, then $\phi_t(x_0) <_r \phi_t(y_0)$.*

PROOF. For $m = 1, 2, \ldots$, let $\phi_t^m(x)$ be the flow corresponding to

$$x' = f(x) + (1/m)e$$

where $e = (1, 1, \ldots, 1)$. Suppose that $x_0 \leq y_0$, $t > 0$ and both $\phi_t(x_0)$ and $\phi_t(y_0)$ are defined. Then (Hale(1980), Chapt.1, Lemma 3.1), $\phi_s^m(y_0 + e/m)$ is defined on $0 \leq s \leq t$ for all large m, say $m > M$, and

$$\phi_s^m(y_0 + e/m) \to \phi_s(y_0)$$

as $m \to \infty$, uniformly in $s \in [0, t]$. We claim that

$$\phi_s(x_0) \ll \phi_s^m(y_0 + e/m) \tag{1.2}$$

holds for $0 \leq s \leq t$, for all $m > M$. Fix $m > M$. Since $x_0 = \phi_0(x_0) \ll y_0 + e/m = \phi_0^m(y_0 + e/m)$, the claim holds for small s by continuity. If the claim were false, then there exists t_0, satisfying $0 < t_0 \leq t$, $\phi_s(x_0) \ll \phi_s^m(y_0 + e/m)$ on $0 \leq s < t_0$, and an index i such that $\phi_{t_0}(x_0)_i = \phi_{t_0}^m(y_0 + e/m)_i$, where the subscript i denotes the i-th component. Also, it follows that

$$\frac{d}{ds}|_{s=t_0}\phi_s(x_0)_i \geq \frac{d}{ds}|_{s=t_0}\phi_s^m(y_0 + e/m)_i.$$

However, as $\phi_{t_0}(x_0)_j \leq \phi_{t_0}^m(y_0 + e/m)_j$ for $j \neq i$, the type K condition implies that

$$f_i(\phi_{t_0}(x_0)) \leq f_i(\phi_{t_0}^m(y_0 + e/m)) < f_i(\phi_{t_0}^m(y_0 + e/m)) + 1/m,$$

or

$$\frac{d}{ds}|_{s=t_0}\phi_s(x_0)_i < \frac{d}{ds}|_{s=t_0}\phi_s^m(y_0 + e/m)_i.$$

This contradiction proves the claim. Taking the limit $m \to \infty$ in (1.2) results in $\phi_t(x_0) \leq \phi_t(y_0)$.

If $x_0 < y_0$ then $\phi_t(x_0) \leq \phi_t(y_0)$ and equality does not hold since ϕ_t is one-to-one. Therefore, $\phi_t(x_0) < \phi_t(y_0)$.

If $x_0 \ll y_0$ then ϕ_t maps the set $[x_0, y_0] \cap D$ into the set $[\phi_t(x_0), \phi_t(y_0)]$ by the first part of the proof. Since ϕ_t is a homeomorphism on D and the former set has nonempty interior, so must the latter set. This holds if and only if $\phi_t(x_0) \ll \phi_t(y_0)$. □

The type K condition is most easily identifiable from the sign structure of the Jacobian matrix of the vector field on suitable domains. The next remark describes this structure.

REMARK 1.1. The type K condition can be expressed in terms of the partial derivatives of f on suitable domains. We say that D is *p-convex* if $tx + (1-t)y \in D$ for all $t \in [0, 1]$ whenever $x, y \in D$ and $x \leq y$. If D is a convex set then it is also p-convex. If D is a p-convex subset of $\mathbb{R}^n$ and

$$\frac{\partial f_i}{\partial x_j}(x) \geq 0, \qquad i \neq j, \quad x \in D, \tag{1.3}$$

holds then the fundamental theorem of calculus, implies that f is of type K in D. In fact, if $a \leq b$ and $a_i = b_i$, then

$$f_i(b) - f_i(a) = \int_0^1 \sum_{j \neq i} \frac{\partial f_i}{\partial x_j}(a + r(b-a))(b_j - a_j) dr \geq 0,$$

by (1.3).

Some extensions of Proposition 1.1 are described in the following remarks.

REMARK 1.2. Proposition 1.1 has an obvious analog for nonautonomous systems. For example, suppose that $f(t, x)$ and $\frac{\partial f}{\partial x}(t, x)$ are continuous in $\mathbb{R}^+ \times D$ where D is an open subset of $\mathbb{R}^n$ and for each $t \geq 0$, $f(t, \bullet)$ satisfies the type K condition. If $x(t)$ and $y(t)$ are two solutions of

$$x' = f(t, x)$$

on the interval $[a, b]$ satisfying $x(a) <_r y(a)$, then $x(b) <_r y(b)$. Here, $<_r$ stands for one of the three relations $\leq, <, \ll$. The proof of this remark is very similar to the proof of Proposition 1.1 so we do not give it here. □

REMARK 1.3. The previous remark can be applied to the linear system
$$y' = Df(x(t))y$$
where $x(t)$ is a solution of (1.1) defined on $\mathbb{R}^+$ and $Df(x)$ is the Jacobian matrix of f at x provided (1.3) holds. In this case, the function $g(t,y) = Df(x(t))y$ satisfies the hypotheses of Remark 1.2. Consequently, if $y(t)$ is a solution of the linear system satisfying $0 <_r y(0)$, then $0 <_r y(t)$ for $t > 0$. □

REMARK 1.4. In the applications it frequently occurs that the natural domain for (1.1) is a closed set D and we would like for the conclusions of Proposition 1.1 to hold on D. Suppose that f is continuously differentiable on some neighborhood of D and that D is the closure of an open set G on which f is type K. Assume that D is positively invariant for (1.1). Suppose further that whenever $x, y \in D$ satisfy $x < y$, there exist sequences $x_n, y_n \in G$ satisfying $x_n < y_n$ and $x_n \to x$, $y_n \to y$ as $n \to \infty$. Then Proposition 1.1 holds on D. The proof follows from continuity of solutions with respect to initial conditions and the fact that Proposition 1.1 holds on G. □

The system (1.1) is said to be a *cooperative system* if (1.3) holds on the p-convex domain D. It is called a *competitive system* on D if D is p-convex and the inequalities (1.3) are reversed:
$$\frac{\partial f_i}{\partial x_j}(x) \leq 0, \quad i \neq j, \quad x \in D.$$
Observe that if (1.1) is a competitive system with flow ϕ_t then
$$x' = -f(x)$$
is a cooperative system with flow ψ_t where $\psi_t(x) = \phi_{-t}(x)$, and conversely. Therefore, by time reversal, a competitive system becomes a cooperative system and vice-versa. This fact will be exploited repeatedly below.

A cooperative system generates a monotone dynamical system; the order relation $\leq$ is preserved by the forward flow according to Proposition 1.1. A competitive system has the property that the time-reversed flow is monotone; if $x \leq y$ and $t < 0$ then $\phi_t(x) \leq \phi_t(y)$. In particular, if x and y are *unrelated*, that is, neither $x \leq y$ nor $y \leq x$ holds, and if $t > 0$ then $\phi_t(x)$ and $\phi_t(y)$ are not related (for if they were, then application of ϕ_{-t} would lead to x and y being related). Therefore, the forward flow of a competitive system preserves the property of two points being unrelated.

2. Positively Invariant Sets and Monotone Solutions

Cooperative and Competitive systems have certain canonical positively invariant sets that are identified in the next result.

PROPOSITION 2.1. *If (1.1) is cooperative and $<_r$ stands for one of the relations $\leq, <, \ll$, then $P_+ = \{x \in D : 0 <_r f(x)\}$ and $P_- = \{x \in D : f(x) <_r 0\}$ are positively invariant. If $x \in P_+$ ($x \in P_-$), then $\phi_t(x)$ is nondecreasing (nonincreasing) for $t \geq 0$. If, in addition, $\gamma^+(x)$ has compact closure in D, then $\omega(x)$ is an equilibrium. If (1.1) is competitive, then $U_+ = \{x \in D : f_i(x) > 0 \text{ for some } i\}$ and $U_- = \{x \in D : f_i(x) < 0 \text{ for some } i\}$ are positively invariant. Also, $V_+ = \{x \in D :$*

$f_i(x) \geq 0$ for some i} and $V_- = \{x \in D : f_i(x) \leq 0$ for some $i\}$ are positively invariant. $V_+ \cap V_-$ is closed, positively invariant and contains any compact invariant set that contains no equilibria.

PROOF. When (1.1) is cooperative, the positive invariance of the sets P_+, P_- follows directly from Remark 1.3 concerning the linear system

$$y' = Df(x(t))y$$

where $x(t)$ is a solution of (1.1). Indeed, if $x(t)$ is a solution of (1.1), then $y(t) = f(x(t))$ is a solution of the linear system. Since (1.3) holds, $\text{Int}\mathbb{R}_+^n, \mathbb{R}_+^n \setminus \{0\}$ and $\mathbb{R}_+^n$ are positively invariant for the linear system. If K denotes any one of these sets then it follows that $-K$ is also positively invariant for the linear system. Now suppose that (1.1) is competitive, $x_0 \in U_+$ and there is a $s > 0$ such that $\phi_s(x_0) \notin U_+$. Then $y_0 = \phi_s(x_0) \in G = \{z \in D : -f(z) \geq 0\}$. G is positively invariant for the time-reversed cooperative system $x' = -f(x)$ by the paragraph above. The flow for the time-reversed system is $\phi_{-t}(x)$ so $x_0 = \phi_{-s}(y_0) \in G$. This contradiction (to $x_0 \in U_+$) proves that U_+ is positively invariant for the competitive system. Similar arguments show the positive invariance of U_-, V_+, V_-. Consequently, $V_+ \cap V_-$ is positively invariant. If x does not belong to V_+ then $-f(x) \gg 0$ so $-f(\phi_{-t}(x)) \gg 0$ for $t > 0$ by the first paragraph of the proof. Therefore, $\frac{d}{dt}\phi_t(x) = f(\phi_t(x)) \ll 0$ for $t < 0$ so $\phi_t(x)$ is strictly decreasing on $t < 0$. If, in addition, $x \in A$, where A is a compact invariant set, then it follows that $\alpha(x)$ is an equilibrium in A. Since A is assumed to have no equilibria, it follows that $A \subset V_+$. A similar argument shows that $A \subset V_-$. $\square$

According to Proposition 2.1, a solution $\phi_t(x)$ of a competitive system which never meets $V_+ \cap V_-$ must be strictly increasing or strictly decreasing in t. If such a solution generates a positive orbit with compact closure in D, then it converges to equilibrium. Furthermore, any periodic orbit must be contained in $V_+ \cap V_-$.

The most important assertion of Proposition 2.1 is the first one for cooperative systems. The positive invariance of the sets P_+ and P_- means that any bounded solution starting in one of these sets is monotone and therefore must converge to equilibrium. Points of P_+ where $<_r = \leq$ are sub-equilibria and points of P_+ where $<_r = <$ are strict sub-equilibria. Similarly, P_- are super-equilibria or strict super-equilibria. See section 5 of Chapter 2.

Figure 2.1 depicts the positively invariant set $V_+ \cap V_-$ for the two dimensional competitive Lotka-Volterra system.

Reasoning similar to that used in the previous proof shows that solutions of planar competitive or cooperative systems converge monotonically to equilibrium.

THEOREM 2.2. Let (1.1) be a competitive or cooperative system on a domain D in $\mathbb{R}^2$. If $x(t)$ is a solution defined for all $t \geq 0$ ($t \leq 0$), then there exists a $T \geq 0$ such that for each $i = 1, 2$, $x_i(t)$ is monotone on $t \geq T$ ($t \leq -T$). In particular, if $\gamma^+(x(0))$ ($\gamma^-(x(0))$) has compact closure in D, then $\omega(x(0))$ ($\alpha(x(0))$) is a single equilibrium.

PROOF. It suffices to consider only the case that $x(t)$ is defined for $t \geq 0$ since otherwise the corresponding solution of the time-reversed system is defined on $t \geq 0$ and this system is also competitive or cooperative.

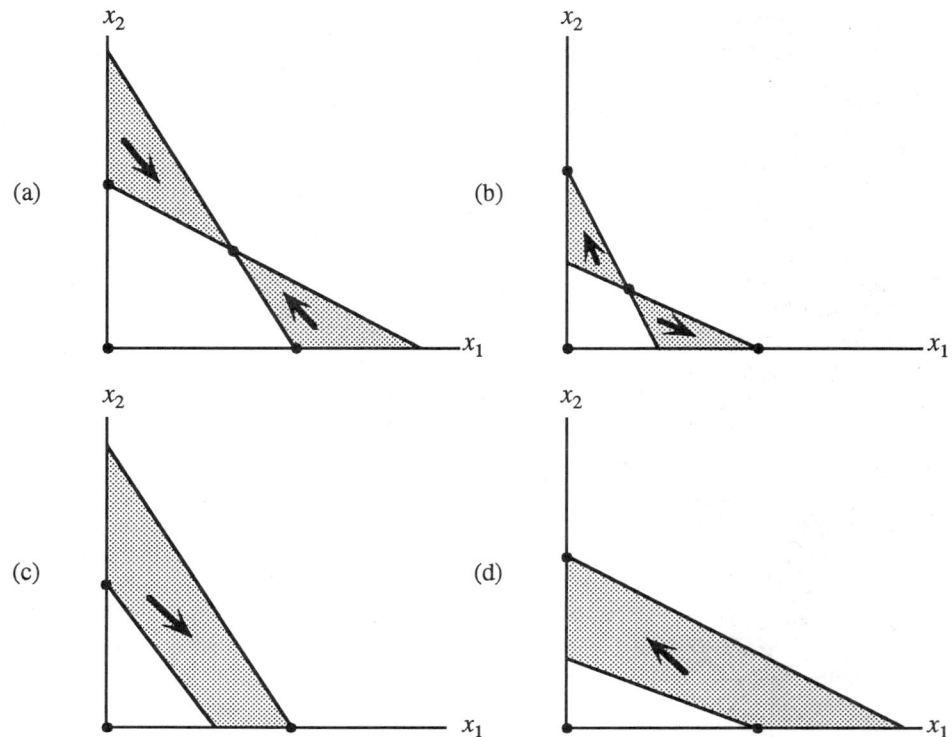

FIGURE 2.1. The (shaded) set $V_+ \cap V_-$ for the competitive Lotka-Volterra system $x_1' = x_1(1 - x_1 - a_{12}x_2)$, $x_2' = \rho x_2(1 - x_2 - a_{21}x_1)$: (a) $a_{12} < 1$, $a_{21} < 1$; (b) $a_{21} > 1$, $a_{12} > 1$; (c) $a_{12} < 1$, $a_{21} > 1$; (d) $a_{12} > 1$, $a_{21} < 1$.

The proof hinges on the identity

$$\frac{d}{dt}(f_1 f_2) = \left(\frac{\partial f_1}{\partial x_1} + \frac{\partial f_2}{\partial x_2}\right)(f_1 f_2) + \frac{\partial f_1}{\partial x_2} f_2^2 + \frac{\partial f_2}{\partial x_1} f_1^2$$

where $\frac{d}{dt}(f_1 f_2)$ is the time derivative of the product of the components f_i of the vector f evaluated along the solution $x(t)$. If (1.1) is cooperative, then $\frac{d}{dt}(f_1 f_2) \geq \left(\frac{\partial f_1}{\partial x_1} + \frac{\partial f_2}{\partial x_2}\right)(f_1 f_2)$; if (1.1) is competitive, then the reverse inequality holds. As a consequence, we show that in the competitive case, the sets $Q_+ = \{x \in D : f_1(x) \geq 0 \text{ and } f_2(x) \leq 0\}$ and $Q_- = \{x \in D : f_1(x) \leq 0 \text{ and } f_2(x) \geq 0\}$ are positively invariant. In the cooperative case, a similar argument shows that $P_+ = \{x \in D : f_1(x) \geq 0 \text{ and } f_2(x) \geq 0\}$ and $P_- = \{x \in D : f_1(x) \leq 0 \text{ and } f_2(x) \leq 0\}$ are positively invariant.

If Q_+ were not positively invariant in the competitive case, then either there exists a point $x_0 \in Q_+$ satisfying $f_1(x_0) = 0$, $f_2(x_0) < 0$ such that $f_1(\varphi_t(x_0)) < 0$ for arbitrarily small positive t or there exists a point $x_0 \in Q_+$ satisfying $f_1(x_0) > 0$, $f_2(x_0) = 0$ such that $f_2(\varphi_t(x_0)) > 0$ for arbitrarily small positive t. Consider the former case as the latter is treated similarly. By the differential inequality satisfied by $f_1(\varphi_t(x_0))f_2(\varphi_t(x_0))$, it follows that $f_1(\varphi_t(x_0))f_2(\varphi_t(x_0)) \leq 0$ for all $t \geq 0$. But this contradicts the conclusion, drawn above, that $f_1(\varphi_t(x_0))f_2(\varphi_t(x_0)) > 0$ for arbitrarily small positive t, proving that Q_+ is positively invariant. Similarly, Q_- is positively invariant.

Obviously, the conclusion of the theorem holds if $x(t) \in Q_+ \cup Q_-$ for some $t \geq 0$. For if $x(t_0) \in Q_+$ then $x_1(t)$ is nondecreasing and $x_2(t)$ is nonincreasing on $t \geq t_0$. If $x(t) \notin Q_+ \cup Q_-$ for all $t \geq 0$, then for each $t \geq 0$, either $f(x(t)) \gg 0$ or $f(x(t)) \ll 0$. In other words, the vector $f(x(t))$ must point into the open first or third quadrants of the plane. Now $\{f(x(t)) : t \geq 0\}$ is connected and therefore either $f(x(t)) \gg 0$ for all $t \geq 0$ or $f(x(t)) \ll 0$ for all $t > 0$. In either case we are done.

The argument in the cooperative case is similar. $\square$

It should be noted that the monotonicity of the flow played no role in the proof of Theorem 2.2 and therefore there is no need to assume that D is p-convex.

3. Main Results

A key result of this chapter is that a compact limit set of a competitive or cooperative system cannot contain two points x_1, x_2 satisfying $x_1 \ll x_2$. If the system is cooperative and the limit set is an omega limit set, then this result follows from the Nonordering of Limit Sets from Chapter 1. The proof of the more general result is based on the next lemma. First, some notation is required. Let $x(t)$ be a solution of the system (1.1) on an interval I. A subinterval $[a,b]$ of I is called a *rising interval* if $x(a) < x(b)$ and a *falling interval* if $x(b) < x(a)$.

LEMMA 3.1. *Let $x(t)$ be a solution of the cooperative system (1.1) on the interval I. Then $x(t)$ cannot have a rising interval and a falling interval that are disjoint.*

PROOF. The most important observation is that if $[a,b]$ is a rising (falling) interval contained in I and if $s > 0$ is such that $[a+s, b+s]$ is contained in I, then it is also a rising (falling) interval in I. In fact, if $x(a) < x(b)$, then by Proposition 1.1, $x(a+s) = \phi_s(x(a)) < \phi_s(x(b)) = x(b+s)$. Consequently, rising and falling intervals remain such under right translation.

Suppose that I contains the falling interval $[a, r]$ and the rising interval $[s, b]$ and suppose that $a < r < s < b$. The other case can be treated similarly. Let $A = \{t \in [s, b] : x(t) \leq x(s)\}$ and $s' = \sup A$. Then $s \leq s' < b$ and $[s', b]$ is a rising interval that contains no falling interval $[s', \eta]$ for any $\eta \in (s', b]$. Redefine $s = s'$ so that the rising interval $[s, b]$ has the above property. A contradiction will be reached in each of the two cases $r - a \leq b - s$ and $r - a > b - s$.

If $r - a \leq b - s$ then $[s, s+r-a]$ is a translate to the right of $[a, r]$, which is contained in I, and therefore is a falling interval. As $s + r - a \leq b$, this contradicts that $[s, b]$ contains no such falling interval.

If $r - a > b - s$ then $a < a + b - r < s < b$ so $[a+b-r, b]$ is a translate to the right of $[a, r]$, contained in I, so it is a falling interval. It follows that $x(s) < x(b) < x(a+b-r)$. Let $c = \sup\{t \in [a+b-r, s] : x(b) \leq x(t)\}$. Then $c < s < b$ and $x(b) \leq x(c)$ so $[c, s]$ is a falling interval adjacent to the rising interval $[s, b]$. If $s - c \leq b - s$ then $[s, 2s - c]$ is a translate to the right of the falling interval $[c, s]$, contained in $[s, b]$, and therefore is a falling interval. But this contradicts the earlier argument that $[s, b]$ contains no such interval. If $s - c > b - s$ then $c < c + b - s < s$ and $[c+b-s, b]$ is a right translate of the falling interval $[c, s]$

so it is a falling interval. Therefore $x(b) < x(c+b-s)$ and this contradicts the definition of c. $\square$

The next result is the key tool for the main results of this chapter.

THEOREM 3.2. *A compact limit set of a competitive or cooperative system cannot contain two points related by $\ll$.*

PROOF. By time reversal if necessary, we can suppose that (1.1) is cooperative. The proof will be given in case that the limit set L is an alpha limit set, $\alpha(x_0)$. The reader should have no problem adapting the argument given in this case to the case that L is an omega limit set. Suppose that $L = \alpha(x_0)$ contains points x_1 and x_2 satisfying $x_1 \ll x_2$. Let $x(t) = \phi_t(x_0)$ for $t \leq 0$. Since $\{x : x_1 \ll x\}$ is an open neighborhood of x_2 and the latter belongs to L, there exists $t_1 < 0$ such that $x_1 \ll x(t_1)$. As $\{x : x \ll x(t_1)\}$ is an open neighborhood of x_1 and the latter belongs to L, there exists t_2 satisfying $t_2 < t_1$ and $x(t_2) \ll x(t_1)$. Continuing in this way, we find $t_3 < t_2$ such that $x(t_3) \ll x_2$ and $t_4 < t_3$ such that $x(t_3) \ll x(t_4)$. Therefore, the interval $I = [t_4, t_1]$ contains the falling interval $[t_4, t_3]$ and the rising interval $[t_2, t_1]$ and these intervals are disjoint. This contradicts Lemma 3.1, proving the Theorem in this case. $\square$

A periodic orbit γ of a competitive or cooperative system is a compact limit set and consequently it cannot contain two points related by $\ll$. Notice that this fact precludes the existence of periodic orbits of competitive or cooperative systems in $\mathbb{R}^2$ since any Jordan curve in the plane necessarily contains two points related by $\ll$. The following sharper result concerning periodic orbits will be required later.

PROPOSITION 3.3. *Let γ be a nontrivial periodic orbit of a competitive or cooperative system. Then γ cannot contain two points that are related by $<$.*

PROOF. We can assume that the system is cooperative. Suppose that $y_i \in \gamma$, $i = 1, 2$ satisfy $y_1 < y_2$. Let $T > 0$ be the minimal period of the solution $\phi_t(y_2) = x(t)$. There exists $\tau \in (0, T)$ such that $x(\tau) = y_1 < y_2 = x(0)$. Therefore, $[0, \tau]$ is a falling interval for $x(t)$. By the periodicity of $x(t)$, $x(\tau + T) = y_1 < y_2 = x(2T)$ so $[\tau + T, 2T]$ is a rising interval disjoint from $[0, \tau]$. This contradicts Lemma 3.1. $\square$

The next result is the main result of this Chapter. The following definitions will make it easier to state.

Let A be an invariant set for (1.1) with flow ϕ_t and let B be an invariant set for the system
$$y' = F(y)$$
with flow ψ_t. We say that the flow ϕ_t on A is *topologically equivalent* to the flow ψ_t on B provided there is a homeomorphism $Q : A \to B$ such that $Q(\phi_t(x)) = \psi_t(Q(x))$ for all $x \in A$ and all $t \in \mathbb{R}$. The relationship of topological equivalence is one of several equivalence relations on the set of all flows that say, roughly, that the dynamics of the two flows are the same. A system of differential equations $y' = F(y)$, defined on $\mathbb{R}^k$, is said to be *Lipschitz* if F is Lipschitz. That is, there exists $K > 0$ such that $|F(y_1) - F(y_2)| \leq K|y_1 - y_2|$ for all $y_1, y_2 \in \mathbb{R}^k$.

THEOREM 3.4. *The flow on a compact limit set of a competitive or cooperative system in $\mathbb{R}^n$ is topologically equivalent to a flow on a compact invariant set of a Lipschitz system of differential equations in $\mathbb{R}^{n-1}$.*

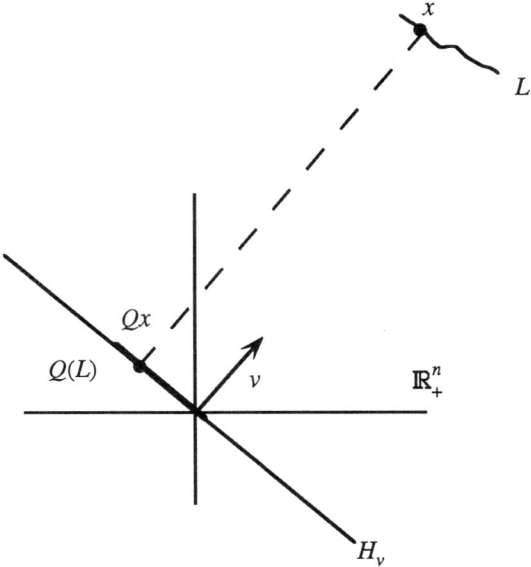

FIGURE 3.1

PROOF. Let L be the limit set. Let v be a unit vector satisfying $0 \ll v$ and let H_v be the hyperplane orthogonal to v. H_v consists of vectors x such that $x \cdot v = 0$, where "$\cdot$" is the standard dot (or scalar product) in $\mathbb{R}^n$. Let Q be the orthogonal projection onto H_v, that is $Qx = x - (x \cdot v)v$. See **Figure 3.1**. By Theorem 3.2, Q is one-to-one on L (this could fail only if L contains two points that are related by $\ll$). Therefore, Q_L, the restriction of Q to L, is a Lipschitz homeomorphism of L onto a compact subset of H_v. We argue by contradiction to establish the existence of $m > 0$ such that $|Q_L x_1 - Q_L x_2| \geq m|x_1 - x_2|$ whenever $x_1 \neq x_2$ are points of L. If this were false, then there exists sequences $x_n, y_n \in L$, $x_n \neq y_n$ such that

$$\frac{|Q(x_n) - Q(y_n)|}{|x_n - y_n|} \to 0$$

as $n \to \infty$. Equivalently,

$$\frac{|(x_n - y_n) - v[v \cdot (x_n - y_n)]|}{|x_n - y_n|} \to 0,$$

or,

$$|w_n - v(v \cdot w_n)| \to 0$$

as $n \to \infty$ where

$$w_n = \frac{x_n - y_n}{|x_n - y_n|}.$$

We can assume that $w_n \to w$ as $n \to \infty$ where $|w| = 1$. Then, $w = v(v \cdot w)$ and therefore, $(v \cdot w)^2 = 1$ so $w = \pm v$. But then

$$\frac{x_n - y_n}{|x_n - y_n|} \to \pm v$$

as $n \to \infty$ and this implies that $x_n \ll y_n$ or $y_n \ll x_n$ for all large n, contradicting Theorem 3.2. Therefore, Q_L^{-1} is Lipschitz on $Q(L)$. Since L is a limit set, it is an invariant set for (1.1). It follows that the dynamical system restricted to L can be modeled on a dynamical system in H_v. In fact, if $y \in Q(L)$ then $y = Q_L(x)$ for a unique $x \in L$ and $\psi_t(y) \equiv Q_L(\phi_t(x))$ is a dynamical system on $Q(L)$ generated by the vector field

$$F(y) = Q_L(f(Q_L^{-1}(y)))$$

on $Q(L)$. According to McShane(1934), a Lipschitz vector field on an arbitrary subset of H_v can be extended to a Lipschitz vector field on all of H_v, preserving the Lipschitz constant. It follows that F can be extended to all of H_v as a Lipschitz vector field. It is easy to see that $Q(L)$ is an invariant set for the latter vector field. We have established the topological equivalence of the flow ϕ_t on L with the flow ψ_t on $Q(L)$. $Q(L)$ is a compact invariant set for the $(n-1)$-dimensional dynamical system on H_v generated by the extended vector field. $\square$

As a consequence of Theorem 3.4, the flow on a compact limit set, L, of a competitive or cooperative system shares common dynamical properties with the flow of a system of differential equations in one less dimension, restricted to a compact invariant set, $Q(L)$. However, notice that $Q(L)$ need not be a limit set of any orbit of the Lipschitz vector field. Obviously, $Q(L)$ is connected since L is connected.

A dynamical property possessed by all limit sets (or, more accurately, the flow restricted to the limit set) is that of *chain recurrence* Conley(1972,1978). The definition is as follows. Let A be a compact invariant set for the flow ψ_t. Given two points z and y in A and positive numbers ϵ and t, an (ϵ, t)-chain from z to y in A is an ordered set

$$\{z = x_1, x_2, \ldots, x_{n+1} = y; t_1, t_2, \ldots, t_n\} \tag{3.1}$$

of points $x_i \in A$ and times $t_i \geq t$ such that

$$|\psi_{t_i}(x_i) - x_{i+1}| < \epsilon, \qquad i = 1, 2, \ldots, n. \tag{3.2}$$

Figure 3.2 depicts such a chain. A is said to be chain recurrent if for every $z \in A$ and for every $\epsilon > 0$ and $t > 0$, there is an (ϵ, t)-chain from z to z in A. In Theorem 1 of the appendix, we prove that alpha and omega limit sets have the property of chain recurrence. It is easy to see that the flow ψ_t on the compact invariant space $Q(L)$ is chain recurrent since it is topologically equivalent to the flow ϕ_t on the chain recurrent limit set L.

4. Three Dimensional Systems

The Poincaré-Bendixson Theorem for three dimensional cooperative and competitive systems, proved below, is the most notable consequence of Theorem 3.4. However, it is harder to apply than its planar analog because it is more difficult to exclude that no equilibrium belongs to the limit set for a three dimensional system.

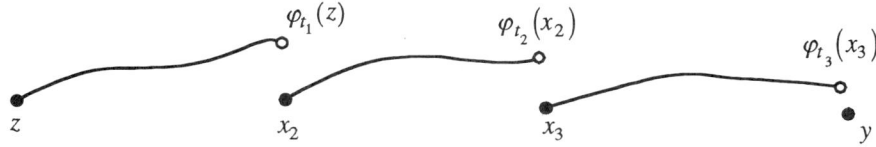

FIGURE 3.2. An (ϵ, t)-chain

Theorem 4.2 of this section gives more easily checked conditions for most solutions to approach a nontrivial periodic orbit when there is only a single unstable equilibrium. The remainder of the section is devoted to describing the sense in which the existence of a periodic orbit implies the existence of a corresponding equilibrium inside it. Essentially, there is a topological 2-sphere, containing the periodic orbit as an "equator", that bounds a positively invariant ball. This ball must contain at least one equilibrium.

THEOREM 4.1. *A compact limit set of a competitive or cooperative system in $\mathbb{R}^3$ that contains no equilibrium points is a periodic orbit.*

PROOF. Let L be the limit set. By Theorem 3.4, the flow ϕ_t on L is topologically equivalent to the flow ψ_t, generated by a Lipschitz planar vector field, restricted to the compact, connected, chain recurrent invariant set $Q(L)$. Since L contains no equilibria neither does $Q(L)$. The Poincaré-Bendixson Theorem implies that $Q(L)$ consists of periodic orbits and, possibly, entire orbits whose omega and alpha limit sets are periodic orbits contained in $Q(L)$. The chain recurrence of the flow ψ_t on $Q(L)$ will be exploited to show that $Q(L)$ consists entirely of periodic orbits. Let $z \in Q(L)$ and suppose that z does not belong to a periodic orbit. Then $\omega(z)$ and $\alpha(z)$ are distinct periodic orbits in $Q(L)$. Let $\omega(z) = \gamma$ and suppose for definiteness that z belongs to the interior component, D, of $\mathbb{R}^2 \setminus \gamma$ so that $\psi_t(z)$ spirals toward γ in D. The other case is treated similarly. Then γ is asymptotically stable relative to D. Standard arguments using transversals (see **Figure 4.1**) imply the existence of compact, positively invariant neighborhoods U_1 and U_2 of γ in D such that $U_2 \subset \text{Int}_D U_1$, $z \notin U_1$ and there exists $t_0 > 0$ for which $\psi_t(U_1) \subset U_2$ for $t \geq t_0$. Let $\epsilon > 0$ be such that the 2ϵ-neighborhood of U_2 in D is contained in U_1. Choose t_0 larger if necessary such that $\psi_t(z) \in U_2$ for $t \geq t_0$. This can be done since $\omega(z) = \gamma$. Then any (ϵ, t_0)-chain (3.1) in $Q(L)$ beginning at $x_1 = z$ satisfies $\psi_{t_1}(x_1) \in U_2$ and, by (3.2) and the fact that the 2ϵ-neighborhood of U_2 is contained in U_1, $x_2 \in U_1$. As $t_2 > t_0$, it then follows that $\psi_{t_2}(x_2) \in U_2$ and (3.2) again implies that $x_3 \in U_1$. Continuing this argument, it is evident that the (ϵ, t_0)-chain cannot return to z. There can be no (ϵ, t_0)-chain in $Q(L)$ from z to z and therefore we have contradicted that $Q(L)$ is chain recurrent. Consequently, every orbit of $Q(L)$ is periodic. Since $Q(L)$ is connected, it is either a single periodic orbit or an annulus consisting of periodic orbits. It follows that L is either a single periodic orbit or a cylinder of periodic orbits.

To complete the proof we must rule out the possibility that $Q(L)$ consists of an annulus of periodic orbits. We can assume that the system is cooperative. The argument will be separated into two cases: $L = \omega(x)$ or $L = \alpha(x)$.

If $L = \omega(x)$ consists of more than one periodic orbit then $Q(L)$ is an annulus of periodic orbits in the plane containing an open subset O. Then there exists

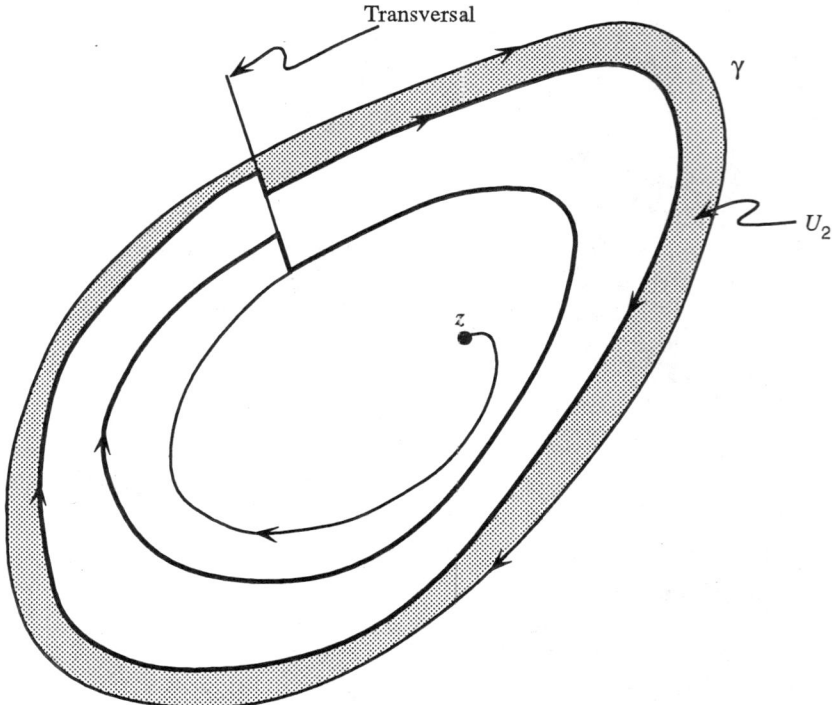

FIGURE 4.1

$t_0 > 0$ such that $Q(\phi_{t_0}(x)) \in O$. Let y be the unique point of L such that $Q(y) = Q(\phi_{t_0}(x))$. $y = \phi_{t_0}(x)$ cannot hold since this would imply that L is a single periodic orbit so it follows that either $y \ll \phi_{t_0}(x)$ or $\phi_{t_0}(x) \ll y$. Suppose that the latter holds, the argument is similar in the other case. Then there exists $t_1 > t_0$ such that $\phi_{t_1}(x)$ is so near y that $\phi_{t_0}(x) \ll \phi_{t_1}(x)$. But then the Convergence Criterion from Chapter 1 implies that $\phi_t(x)$ converges to equilibrium, a contradiction to our assumption that L contains no equilibria. This proves the theorem in this case.

If $L = \alpha(x)$ and $Q(L)$ consists of an annulus of periodic orbits, let $C \subset L$ be a periodic orbit such that $Q(L)$ contains C in its interior. $Q(C)$ separates $Q(L)$ into two components. Fix a and b in $L \setminus C$ such that $Q(a)$ and $Q(b)$ belong to different components of $Q(L) \setminus Q(C)$. Since $\phi_t(x)$ repeatedly visits every neighborhood of a and b as $t \to -\infty$, $Q(\phi_t(x))$ must cross $Q(C)$ at a sequence of times $t_k \to -\infty$. Therefore, there exist $z_k \in C$ such that $Q(z_k) = Q(\phi_{t_k}(x))$ and consequently, as in the previous case, either $z_k \ll \phi_{t_k}(x)$ or $\phi_{t_k}(x) \ll z_k$ holds for each k. Passing to a subsequence, we can assume that either $z_k \ll \phi_{t_k}(x)$ holds for all k or $\phi_{t_k}(x) \ll z_k$ holds for all k. Assume the latter, the argument is essentially the same in the other case. We claim that for every $s < 0$ there is a point $w \in C$ such that $\phi_s(x) > w$.

For if $t_k < s$ then
$$\phi_s(x) = \phi_{s-t_k} \circ \phi_{t_k}(x) < \phi_{s-t_k}(z_k) \in C.$$

If $y \in L$ then $\phi_{s_n}(x) \to y$ for some sequence $s_n \to -\infty$. By the claim, there exists $w_n \in C$ such that $\phi_{s_n}(x) > w_n$. Passing to a subsequence if necessary, we can assume that $w_n \to w \in C$ and $y \geq w$. Therefore, every point of L is related by $\leq$ to some point of C.

The same reasoning applies to every periodic orbit $C' \subset L$ for which $Q(C')$ belongs to the interior of $Q(L)$: either every point of L is $\leq$ some point of C' or every point of L is $\geq$ some point of C'. Since there are three different periodic orbits in L whose projections are contained in the interior of $Q(L)$, there will be two of them for which the same inequality holds between points of L and points of the orbit. Consider the case that there are two periodic orbits C_1 and C_2 such that every point of L is $\leq$ some point of C_1 and $\leq$ some point of C_2. The case that the opposite relations hold is treated similarly. If $u \in C_1$ then it belongs to L so we can find $w \in C_2$ such that $u < w$ (equality can't hold since the points belong to different periodic orbits). But $w \in L$ so we can find $z \in C_1$ such that $w < z$. Consequently, $u, z \in C_1$ satisfy $u < z$, a contradiction to Proposition 3.3. This completes the proof. $\square$

A cooperative system of differential equations cannot have an attracting periodic orbit by Theorem 1.2.2 but a competitive system can. The next result gives sufficient conditions for most orbits of a three dimensional competitive system to be asymptotic to a periodic orbit. Recall that an equilibrium p is said to be *hyperbolic* if the Jacobian matrix of f at p has no purely imaginary eigenvalues. We use the notation $W^s(p)$ for the *stable manifold* of p (See e.g. Hale(1980)).

THEOREM 4.2. *Let (1.1) be a competitive system in $D \subset \mathbb{R}^3$ and suppose that D contains a unique equilibrium point p which is hyperbolic. Suppose further that $W^s(p)$ is one-dimensional and tangent at p to a vector $v \gg 0$. If $q \in D \setminus W^s(p)$ and $\gamma^+(q)$ has compact closure in D then $\omega(q)$ is a nontrivial periodic orbit.*

PROOF. The result will follow from Theorem 4.1 provided $p \notin \omega(q)$. Clearly, $\omega(q) \neq p$ since $q \notin W^s(p)$. If $p \in \omega(q)$ then the Butler-McGehee Lemma (Butler and Waltman (1986)) implies that $\omega(q)$ contains a point y on $W^s(p)$ different from p. By the invariance of $\omega(q)$ we may assume that y is a point of the local stable manifold, arbitrarily close to p. Since $W^s(p)$ is tangent at p to a positive vector, it can be assumed that either $y \ll p$ or $p \ll y$. In either case we have contradicted Theorem 3.2 since both p and y belong to $\omega(q)$. $\square$

The requirement that $W^s(p)$ be tangent to a positive vector can be verified by showing that the Jacobian matrix of f at p is irreducible. This follows from the Perron-Frobenius Theorem (Theorem 4.3.1). Indeed, the Jacobian matrix of f at p, $Df(p)$, has the property that $A = -Df(p) + sI \geq 0$ for all large values of s, where I denotes the identity matrix and the inequality signifies that each entry of the matrix is nonnegative. If $Df(p)$ is irreducible then so is A and consequently, by the Perron-Frobenius Theorem, A has a positive eigenvector v corresponding to the eigenvalue $r > 0$ with maximum modulus. Then, v is an eigenvector for $Df(p)$ corresponding to the eigenvalue $s - r$. The eigenvalues λ of $Df(p)$ give rise to the eigenvalues $s - \lambda$ of A, so, as r is the eigenvalue of A with maximal real part, $s - r$

is the eigenvalue of $Df(p)$ of minimal real part. The hypotheses of Theorem 4.2 imply that $s - r < 0$ and the corresponding eigenvector is v. See Corollary 4.3.2 for a similar argument.

A remarkable fact about three dimensional competitive or cooperative systems on suitable domains is that the existence of a periodic orbit implies the existence of an equilibrium point inside a certain closed ball having the periodic orbit on its boundary. The next result makes this assertion precise. The proof is quite long and tedious, so it might be skipped on first reading. We can assume the system is competitive. Let γ denote the periodic orbit and assume that there exist p, q with $p \ll q$ such that
$$\gamma \subset [p, q] \subset D. \tag{4.1}$$
Define
$$K = \{x \in \mathbb{R}^3 : x \text{ is not related to any point } y \in \gamma\}$$
$$= (\gamma + \mathbb{R}^3_+)^c \cap (\gamma - \mathbb{R}^3_+)^c.$$
Here we use the notation B^c for the complement of the subset B in $\mathbb{R}^3$. Another way to define K is as
$$K^c = \cup_{y \in \gamma}[(y + \mathbb{R}^3_+) \cup (y - \mathbb{R}^3_+)].$$

PROPOSITION 4.3. *Let γ be a non-trivial periodic orbit of a competitive system in $D \subset \mathbb{R}^3$ and suppose that (4.1) holds. Then K is an open subset of $\mathbb{R}^3$ consisting of two connected components, one bounded and one unbounded. The bounded component, $K(\gamma)$, is homeomorphic to the open unit ball in $\mathbb{R}^3$. $K(\gamma) \subset [p, q]$, is positively invariant and its closure contains an equilibrium.*

PROOF. That K is open is a consequence of the fact that $\gamma + \mathbb{R}^3_+$ and $\gamma - \mathbb{R}^3_+$ are closed. We show that $K \cap D$ is positively invariant. If $x \in K \cap D$, $y \in \gamma$ and $t > 0$ then $\phi_{-t}(y) \in \gamma$ so x is not related to it. Since the forward flow of a competitive system preserves the property of being unrelated, $\phi_t(x)$ is unrelated to y. Therefore, $\phi_t(x) \in K \cap D$.

As in the proof of Theorem 3.4, let $v \gg 0$ be a unit vector, H_v be the hyperplane orthogonal to v and Q be the orthogonal projection onto H_v along v. Q is one-to-one on γ so $Q(\gamma)$ is a Jordan curve in H_v. Let H_i and H_e denote the interior and exterior components of $H_v \setminus Q(\gamma)$. If $x \in Q^{-1}(Q(\gamma))$ then $Q(x) = Q(y)$ for some $y \in \gamma$ and therefore either $x = y$, $x \ll y$ or $y \ll x$. In any case, $x \notin K$. Hence,
$$K = (K \cap Q^{-1}(H_i)) \cup (K \cap Q^{-1}(H_e)).$$
Set $K(\gamma) = K \cap Q^{-1}(H_i)$. There exists $M > 0$ such that for all $z \in H_i$, $z + tv \gg \gamma$ for all $t > M$ and $z + tv \ll \gamma$ for all $t < -M$ since $v \gg 0$. It follows that $\{t \in \mathbb{R} : z + tv \leq y \text{ for some } y \in \gamma\} = (-\infty, t_-(z)]$ where $t_-(z) \leq M$ and there exists $y_-(z) \in \gamma$ such that
$$z + t_-(z)v \leq y_-(z). \tag{4.2}$$
The maximality of $t_-(z)$ implies that $y_-(z)$ and $z + t_-(z)v$ share at least one common component. Thus $-M \leq t_-(z)$. Similarly, $\{t \in \mathbb{R} : z + tv \geq y \text{ for some } y \in \gamma\} = [t_+(z), \infty)$ where $t_+(z) \geq -M$ and there exists $y_+(z) \in \gamma$ such that
$$z + t_+(z)v \geq y_+(z). \tag{4.3}$$
The minimality of $t_+(z)$ implies that $y_+(z)$ and $z+t_+(z)v$ share at least one common component. Thus $t_+(z) \leq M$. The inequalities (4.2) and (4.3) lead to $(t_+(z) -$

$t_-(z))v \geq y_+(z) - y_-(z)$ or $y_-(z) \geq y_+(z) + (t_-(z) - t_+(z))v$. If $t_-(z) > t_+(z)$ then $y_-(z) \gg y_+(z)$ which contradicts Proposition 3.2. If $t_-(z) = t_+(z)$ then $y_-(z) \geq y_+(z)$ and by Proposition 3.3, it must be the case that $y_-(z) = y_+(z)$. Therefore, $y_+(z) \leq z + t_+(z)v = z + t_-(z)v \leq y_-(z)$ and equality must hold so $z + t_+(z)v = y_+(z)$ or $z \in Q(\gamma)$. This contradiction proves that $t_-(z) < t_+(z)$. It follows that $z + tv \in K$ for $t_-(z) < t < t_+(z)$. Hence

$$K(\gamma) = \{z + tv : z \in H_i, t_-(z) < t < t_+(z)\}.$$

Now we show that the maps $z \to z + t_-(z)v$ and $z \to z + t_+(z)v$ are continuous on H_i. Suppose $z_n, z \in H_i$ and $z_n \to z$ as $n \to \infty$. As $-M \leq t_-(z_n) \leq M$, we can select a convergent subsequence $t_k = t_-(z_{n_k})$ such that $t_k \to \bar{t}$. By passing to a subsequence if necessary, we can also suppose that the corresponding subsequence $y_k = y_-(z_{n_k})$ converges to $\bar{y} \in \gamma$. Since $z_n + t_-(z_n)v \leq y_-(z_n)$ it follows that $z + \bar{t}v \leq \bar{y}$. But $z + t_-(z)v \leq y_-(z)$ and by the maximality of $t_-(z)$ it follows that $\bar{t} \leq t_-(z)$. If strict inequality holds, then, since $v \gg 0$, $z_{n_k} + t_-(z_{n_k})v \ll y_-(z)$ for all large k, contradicting the maximality of $t_-(z_{n_k})$ for these k. Consequently, $\bar{t} = t_-(z)$ and every convergent subsequence of $t_-(z_n)$ must converge to $t_-(z)$ and so $t_-(z_n) \to t_-(z)$ as $n \to \infty$. This establishes the continuity of the map $z \to z + t_-(z)v$ on H_i. Similar arguments show that $t_+(z)$ is continuous. Finally, suppose that $w \in Q(\gamma)$. There is a unique $y \in \gamma$ such that $Q(y) = w$ or, equivalently, $y = w + sv$ for $s = y \cdot v$. Given $\epsilon > 0$ there is a neighborhood U of w in H_v such that $z + (s-\epsilon)v \ll y \ll z + (s+\epsilon)v$ for all $z \in U$. Hence, $s - \epsilon \leq t_-(z) < t_+(z) \leq s + \epsilon$ for all $z \in U \cap H_i$ and therefore, $t_+(z) \to s$ and $t_-(z) \to s$ as $z \to w$ in H_i. Since H_i is homeomorphic to the open unit ball in $\mathbb{R}^2$, these arguments imply that $K(\gamma)$ is homeomorphic to the open unit ball in $\mathbb{R}^3$. Consequently, it is connected. Similar arguments can be applied to H_e to show that

$$K \cap Q^{-1}(H_e) = \{z + tv : z \in H_e, \ t_-(z) < t < t_+(z)\}$$

and that the maps $z \to z + t_-(z)v, z + t_+(z)v$ are continuous. This gives the unbounded component of K.

As $K(\gamma)$ is a connected component of the positively invariant set K, it follows that it is positively invariant. Consequently, its closure is a positively invariant set homeomorphic to the closed unit ball in $\mathbb{R}^3$. It must contain an equilibrium by a standard argument using the Brouwer Fixed Point Theorem (e.g. Hale(1980), Thm.I.8.2).

In order to show that $K(\gamma) \subset [p, q]$, it is convenient to use the same arguments as above only with the projection Q replaced by the projection onto the (x_1, x_2)-plane. Let $\pi : \mathbb{R}^3 \to \mathbb{R}^2$ be defined by $\pi(x_1, x_2, x_3) = (x_1, x_2, 0)$ and set $e_3 = (0, 0, 1)$. Then π is one-to-one on γ by Proposition 3.3 so the image of γ under π is a Jordan curve, $\pi(\gamma)$, in the plane. Let H_i and H_e denote the interior and exterior components of $\mathbb{R}^2 \setminus \pi(\gamma)$. Then

$$K = (K \cap \pi^{-1}(H_i)) \cup (K \cap \pi^{-1}(H_e))$$

since every point of $\pi^{-1}(\pi(\gamma))$ is related to some point of γ and therefore cannot belong to K. Now, the unbounded component of K contains points with arbitrarily large x_1 and x_2 components and it is mapped by π onto an open, connected subset of $H_i \cup H_e$ having $\pi(\gamma)$ as part of its boundary. It follows that this component is projected into H_e by π. On the other hand, $\pi(K(\gamma))$ is an open, bounded,

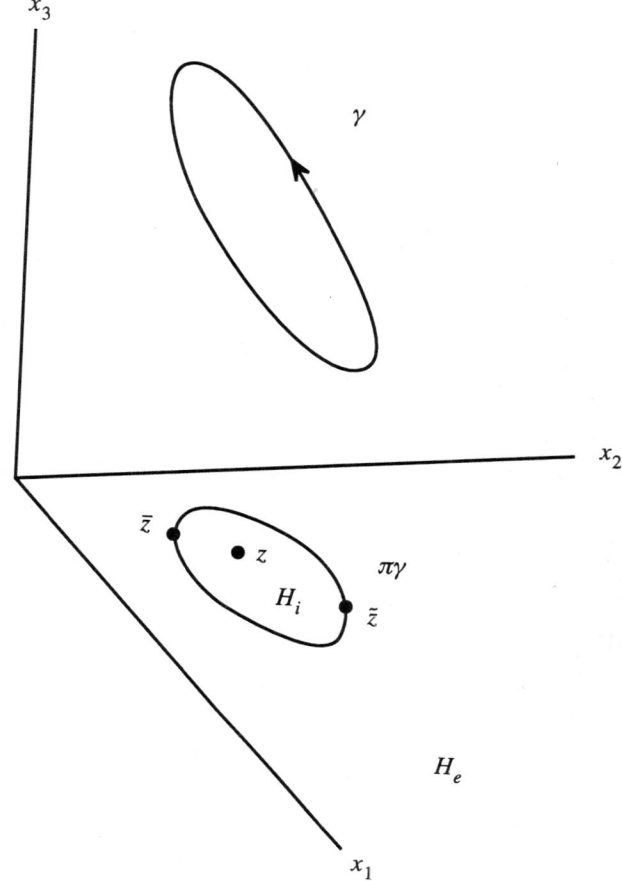

FIGURE 4.2

connected subset of $H_i \cup H_e$ and $\pi(\gamma)$ belongs to its boundary as well. Since $H_i \subset \pi(K)$, as shown below, it follows that $\pi(K(\gamma)) = H_i$. Similarly, $H_e \subset \pi(K)$ so the projection of the unbounded component of K is H_e. Therefore, $K(\gamma) = K \cap \pi^{-1}(H_i)$. We now show that $H_i \subset \pi(K)$. Let $z = (x_1, x_2, 0) \in H_i$. Then, there are points $\bar{z}, \tilde{z} \in \pi(\gamma)$ such that $\bar{z} \leq z \leq \tilde{z}$. See **Figure 4.2**. Let $\bar{z} = \pi(\bar{x})$ and $\tilde{z} = \pi(\tilde{x})$ where $\bar{x}$ and $\tilde{x}$ belong to γ. Consider the set $A(z) = \{t : z + te_3 \leq y, \text{for some } y \in \gamma\}$. It is bounded above since $\max\{y_3 : y \in \gamma\}$ is bounded. If $t \leq \tilde{x}_3$ then $z + te_3 \leq \tilde{x}$ so $t \in A(z)$. It follows that $A(z) = (-\infty, t_-(z)]$ where $t_-(z) \equiv \sup A(z)$ and there exists $y_-(z) \in \gamma$ such that $z + t_-(z)e_3 \leq y_-(z)$. Similar arguments show that $\{t : z + te_3 \geq y \text{ for some } y \in \gamma\} = [t_+(z), \infty)$ and there exists $y_+(z) \in \gamma$ such that $z + t_+(z)e_3 \geq y_+(z)$. Observe that $z \neq \pi(y_-(z)), \pi(y_+(z))$ since the latter belong to $\pi(\gamma)$. Also, $t_-(z) = y_-(z)_3$ and $t_+(z) = y_+(z)_3$. If $t_+(z) < t_-(z)$ then, as $z + t_-(z)e_3 \leq y_-(z)$ and $z + t_+(z)e_3 \geq y_+(z)$, we have $y_-(z) \geq y_+(z) + (t_-(z) - t_+(z))e_3$. Thus, $y_-(z) > y_+(z)$, contradicting Proposition 3.3. If $t_-(z) = t_+(z)$ then $y_+(z) \leq z + t_+(z)e_3 = z + t_-(z)e_3 \leq y_-(z)$ so, by Proposition 3.3, we conclude that $y_-(z) = y_+(z) = z + t_-(z)e_3$. Consequently,

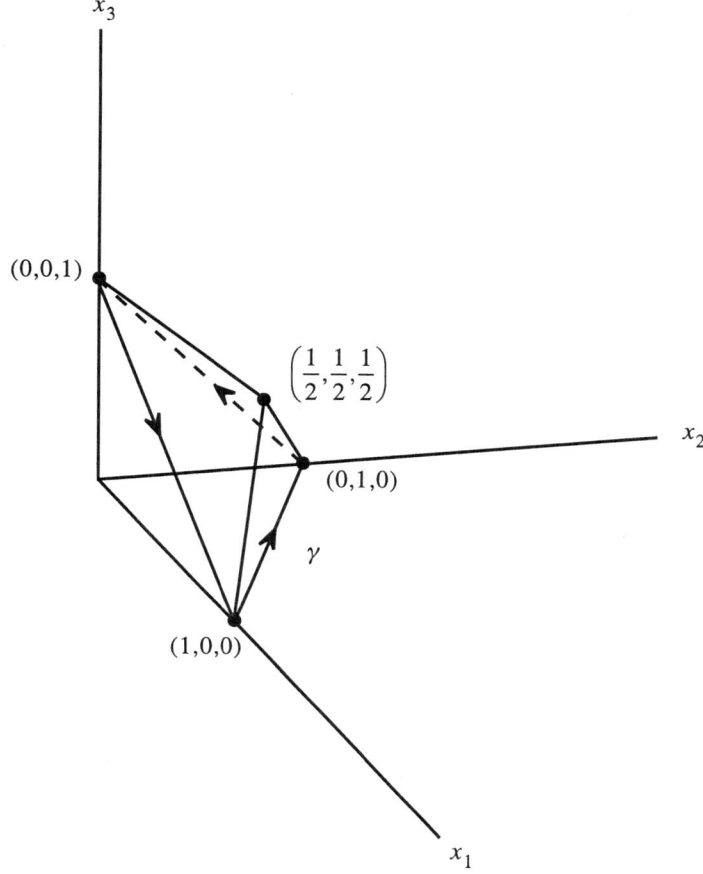

FIGURE 4.3

$z = \pi(y_+(z)) = \pi(y_-(z))$, contradicting the observation above. Therefore, $t_-(z) < t_+(z)$ and $z + te_3 \in K(\gamma)$ for $t_-(z) < t < t_+(z)$. We conclude that $K(\gamma) = \{z + te_3 : z \in H_i, t \in (t_-(z), t_+(z))\}$.

Now we show that $K(\gamma) \subset [p, q]$. If $x \in K(\gamma)$ then $x = z + te_3$ for some $z \in H_i$ and some $t \in (t_-(z), t_+(z))$. Since $\gamma \subset [p, q]$ it follows that $\pi(\gamma) \subset \pi[p, q] = [\pi(p), \pi(q)]$ and therefore, $H_i \subset [\pi(p), \pi(q)]$. Hence, $p_i \leq x_i = z_i \leq q_i$, $i = 1, 2$ for every $x \in K(\gamma)$. Now, $x_3 = t < t_+(z) = y_+(z)_3$ and $y_+(z) \in \gamma$ implies $y_+(z)_3 \leq q_3$ by (3.6), so $x_3 \leq q_3$. Similarly, $x_3 > t_-(z) = y_-(z)_3 \geq p_3$. Therefore, $p \leq x \leq q$ and $K(\gamma) \subset [p, q] \subset D$. □

As the geometry of the set $K(\gamma)$ is difficult to visualize, we offer an (artificial) example where at least it is possible to give an explicit description of $K(\gamma)$. Let γ consist of the three line segments forming the boundary of the standard simplex $S^2 = \{x \in \mathbb{R}^3_+ : x_1 + x_2 + x_3 = 1\}$, oriented as in **Figure 4.3**. Although γ cannot be a periodic orbit, it is a Jordan curve with the property that $x < y$ does not hold for any two points $x, y \in \gamma$. These are the only properties that were used in the proof of Proposition 4.3. Then,

$$K(\gamma) = \{x : x_i > 0, x_1 + x_2 < 1, x_1 + x_3 < 1, x_2 + x_3 < 1\}.$$

This is easy to see. If $x \geq y$ for some $y \in \gamma$, then it follows that $x_i + x_j \geq 1$ for some $i \neq j$, so $x \notin K(\gamma)$. If $x \leq y$ for some $y \in \gamma$ then $x_i \leq 0$ for some i, so $x \notin K(\gamma)$. Therefore, no point of the set above contains a point that is related to a point of γ. On the other hand, if x belongs to any one of the six planes bounding $K(\gamma)$, then it is related to a point of γ. For example, if $x_1 + x_3 = 1$ but each of the five other inequalities are satisfied, then $x \geq y \equiv (x_1, 0, x_3) \in \gamma$.

5. Alternative Cones

It is advantageous to consider other cones besides $\mathbb{R}_+^n$ since by doing so, we can enlarge the class of competitive and cooperative systems. This is the primary motivation for this section. We restrict attention here to the case that the cone is one of the other orthants of $\mathbb{R}^n$. While it is a straightforward exercise to translate the Kamke condition to analogous conditions for other partial orders, the problem of identifying a monotone system, with respect to some partial order, from the vector field becomes more difficult. That is, given a vector field, how are we to determine if it is a monotone system and what is the appropriate partial order. Much of this section is devoted to the identification problem.

We begin with some notation identifying the various orthants. Let $m = (m_1, m_2, \ldots, m_n)$ where $m_i \in \{0, 1\}$ and

$$K_m = \{x \in \mathbb{R}^n : (-1)^{m_i} x_i \geq 0, \ 1 \leq i \leq n\}.$$

K_m is a cone in $\mathbb{R}^n$ and as such, it generates a partial order $\leq_m$ defined by $x \leq_m y$ if and only if $y - x \in K_m$. Equivalently, $x_i \leq y_i$ for those i for which $m_i = 0$ and $y_i \leq x_i$ for those i for which $m_i = 1$. We write $x <_m y$ when $x \leq_m y$ and $x \neq y$, and $x \ll_m y$ when $y - x \in \text{Int} K_m$ or, equivalently, when $x_i < y_i$ for those i for which $m_i = 0$ and $y_i < x_i$ for those i for which $m_i = 1$. Let P be the diagonal matrix defined by $P = \text{diag}[(-1)^{m_1}, (-1)^{m_2}, \ldots, (-1)^{m_n}]$. Observe that $P = P^{-1}$ is an order isomorphism, that is $x \leq_m y$ if and only if $Px \leq Py$. The domain D is said to be p_m-convex if $tx + (1 - t)y \in D$ whenever $x, y \in D$, $0 < t < 1$, and $x \leq_m y$. (1.1) is cooperative with respect to K_m provided D is p_m-convex and

$$(-1)^{m_i + m_j} \frac{\partial f_i}{\partial x_j}(x) \geq 0, \quad i \neq j, \quad x \in D. \tag{5.1}$$

It is competitive with respect to K_m if the reverse inequality holds

$$(-1)^{m_i + m_j} \frac{\partial f_i}{\partial x_j}(x) \leq 0, \quad i \neq j, \quad x \in D. \tag{5.2}$$

If $m = 0$ then we recover the earlier definitions.

PROPOSITION 5.1. *Let D be p_m-convex and f be a continuously differentiable vector field on D such that (5.1) holds. Let $<_r$ denote any one of the relations $\leq_m$, $<_m, \ll_m$. If $x <_r y$, $t > 0$ and $\phi_t(x)$ and $\phi_t(y)$ are defined, then $\phi_t(x) <_r \phi_t(y)$. If (5.2) holds, then similar conclusions are valid for $t < 0$.*

PROOF. PD is p-convex and $g : PD \to \mathbb{R}^n$, defined by $g(y) = Pf(Py)$, satisfies (1.3) if (5.1) holds. The vector field g generates a flow ψ_t defined by $\psi_t(y) = P\phi_t(Py)$. If $x \leq_m y$, it follows that $Px \leq Py$ and therefore, by Proposition 1.1, $\psi_t(Px) \leq \psi_t(Py)$. This implies that $\phi_t(x) \leq_m \phi_t(y)$, as asserted. The other assertions follow similarly. □

Another way to express Proposition 5.1 is to say that (1.1) is cooperative (competitive) with respect to K_m if and only if, in the variable $y = Px$, the resulting system is cooperative (competitive), that is,

$$y' = g(y), \quad g(y) \equiv Pf(Py)$$

is cooperative (competitive), as defined at the beginning of this chapter. It is clear that all the results of this chapter that hold for cooperative (competitive) systems also hold for systems that are cooperative (competitive) with respect to K_m for some m.

Proposition 5.1 suggests an algorithm for determining whether a given system (1.1) is cooperative or competitive in a domain D with respect to one of the cones K_m. It clearly suffices to consider only the cooperative case since the competitive case can be determined by time reversal. The first test is to check that the off-diagonal elements of the Jacobian matrix are *sign-stable* in D. This means that for each $i \neq j$, either (a) $\frac{\partial f_i}{\partial x_j}(x) \geq 0$ for all $x \in D$, or (b) $\frac{\partial f_i}{\partial x_j}(x) \leq 0$ for all $x \in D$. Assuming this test is passed, then the Jacobian matrix must be tested for *sign-symmetry*: $\frac{\partial f_i}{\partial x_j}(x)\frac{\partial f_j}{\partial x_i}(y) \geq 0$ for all $i \neq j$, and all $x, y \in D$. If this test is satisfied, then for each $i < j$ set $s_{ij} = 0$ if $\frac{\partial f_i}{\partial x_j}(x) + \frac{\partial f_j}{\partial x_i}(x) > 0$ for some $x \in D$, $s_{ij} = 1$ if $\frac{\partial f_i}{\partial x_j}(x) + \frac{\partial f_j}{\partial x_i}(x) < 0$ for some $x \in D$, and let $s_{ij} \in \{0, 1\}$ be arbitrary if $\frac{\partial f_i}{\partial x_j}(x) + \frac{\partial f_j}{\partial x_i}(x)$ vanishes identically in D. Now, consider the system of $n(n-1)/2$ linear equations in the n unknowns $m_i \in \{0, 1\}$, given by

$$m_i + m_j = s_{ij} \pmod{2}, \quad i < j, \tag{5.3}$$

Those equations among (5.3) for which s_{ij} is arbitrary may be deleted since they can be satisfied by appropriate choice of s_{ij}. If the remaining equations can be solved for m then (1.1) is cooperative with respect to K_m. It can be seen that if m solves (5.3) then so does $m + 1$.

There is a relatively simple graph theoretic test for determining if a given system is cooperative or competitive in a domain D with respect to some K_m. Assume that the system is sign-stable and sign-symmetric in the domain D. Consider the graph G with vertices $\{1, 2, \ldots, n\}$ where an undirected edge connects vertices i and j if at least one of the partial derivatives $\frac{\partial f_i}{\partial x_j}(x)$ or $\frac{\partial f_j}{\partial x_i}(x)$ does not vanish identically in D. Attach a sign $+$ or $-$ to the edge depending on the sign of one of the partial derivatives at a point where it is nonzero. Then (1.1) is cooperative in D with respect to some K_m if and only if for every closed loop in G the number of edges with $-$ signs is even; it is competitive if every closed loop in G has an odd number of edges with $-$ signs. If the test indicates that the system is cooperative, then the appropriate cone K_m must then be determined by solving (5.3). If the test indicates that the system is competitive, then the time-reversed system is cooperative and the appropriate cone is determined by solving (5.3) after all entries of the Jacobian have been multiplied by -1 to account for time-reversal.

A canonical form for a system that is cooperative with respect to K_m for some m can be obtained by permuting equations and variables in the same way so that m consists of k ones followed by $n - k$ minus ones. This leads to a system for $x = (x_1, x_2) \in \mathbb{R}^k \times \mathbb{R}^{n-k}$ where $x_1 \in \mathbb{R}^k$ and $x_2 \in \mathbb{R}^{n-k}$ of the form

$$x_1' = f_1(x_1, x_2)$$
$$x_2' = f_2(x_1, x_2).$$

The k-by-k ($(n-k)$-by-$(n-k)$) Jacobian $\partial f_1/\partial x_1$ ($\partial f_2/\partial x_2$) has nonnegative off-diagonal entries, while the k-by-$(n-k)$ ($(n-k)$-by-k) Jacobian $\partial f_1/\partial x_2$ ($\partial f_2/\partial x_1$) has nonpositive entries.

A planar competitive system is cooperative with respect to K_m where $m = (0, 1)$ provided that D is p_m-convex. That is, the forward flow is monotone with respect to the partial order generated by the cone $K_m = \{(x_1, x_2) \in \mathbb{R}^2 : x_1 \geq 0, x_2 \leq 0\}$.

6. The Field-Noyes Model

The following example illustrates the use of Theorem 4.2. The differential equations below are a scaled version of the Field-Noyes model of the Belousov-Zhabotinski reaction, due to Murray(1989).

$$\epsilon x' = y - xy + x(1 - qx)$$
$$y' = -y - xy + 2fz \tag{6.1}$$
$$z' = \delta(x - z).$$

The parameters ϵ, q, f, δ are positive and, in the interesting case, $0 < q < 1$. The cube

$$D = \{(x, y, z) : 1 < x < q^{-1}, \frac{2fq}{1+q} < y < \frac{f}{q}, 1 < z < q^{-1}\}$$

is positively invariant since the vector field points strictly inward on the boundary of D. The Jacobian matrix of the vector field at a point (x, y, z) is given by

$$J = \begin{pmatrix} \frac{1-y-2qx}{\epsilon} & \frac{1-x}{\epsilon} & 0 \\ -y & -1-x & 2f \\ \delta & 0 & -\delta \end{pmatrix}.$$

It is immediately apparent that the off-diagonal entries of J are sign-stable and sign-symmetric in D. The incidence graph of J on the vertices x, y, z is given in **Figure 6.1** below. It consists of a single loop with one minus sign on the edge connecting x to y. Therefore, the system is competitive in D. In order to determine the appropriate partial order relation for the time-reversed cooperative system, which has Jacobian $-J$, we need to solve (5.3) where the s_{ij} are determined from $-J$. This leads to the system

$$m_1 + m_2 = 0$$
$$m_1 + m_3 = 1.$$
$$m_2 + m_3 = 1$$

A solution is $m_1 = m_2 = 0, m_3 = 1$. This corresponds to the cone of positive vectors given by

$$K_m = \{(x, y, z) \in \mathbb{R}^3 : x \geq 0, y \geq 0, z \leq 0\}.$$

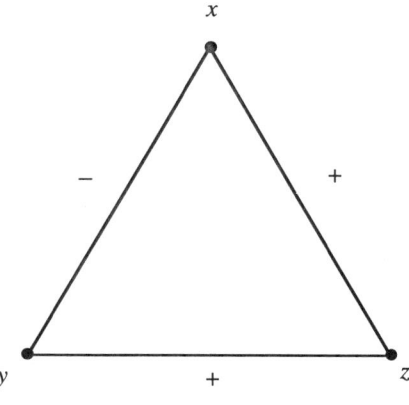

FIGURE 6.1

The reader may have guessed immediately that changing z to $-z$ in (6.1) results in a system satisfying (1.3).

The system (6.1) has a unique equilibrium $p = (x^*, y^*, z^*)$ in D given by

$$x^* = z^*, \quad y^* = \frac{2fx^*}{1+x^*}, \quad q(x^*)^2 + (2f + q - 1)x^* - (2f + 1) = 0,$$

taking the positive root of the quadratic. The determinant of J at p is

$$\text{Det}(J) = -\frac{\delta}{\epsilon}(2y^* + qx^* + q(x^*)^2),$$

where we used the quadratic equation determining x^* to simplify the expression. As the determinant is negative, there are two possibilities if we consider only the case that p is hyperbolic. Either p is asymptotically stable or it is unstable with a one dimensional stable manifold. Both cases can occur, depending on the values of the parameters (see Murray(1989)). Theorem 4.2 can be applied in the second case and leads to the following.

PROPOSITION 6.1. *Suppose that p is hyperbolic and unstable for (6.1). Then the stable manifold of p, $W^s(p)$, is one dimensional and $\omega(q)$ is a nontrivial periodic orbit in D for every point $q \in D \setminus W^s(p)$.*

PROOF. By Theorem 4.2, it is only necessary to check that the one dimensional stable manifold is tangent at p to a positive vector, that is, to a vector in K_m. Since J at p is irreducible, this follows from the discussion following Theorem 4.2, taking into account the order relation generated by K_m. □

Figure 6.2 illustrates the Proposition.

7. Remarks and Discussion

Proposition 1.1 is due to Kamke(1932) and Müller(1926). See Coppel(1965) or Walter(1970) for more general results. Competitive and cooperative systems were brought to the fore by the remarkable series of papers of Hirsch(1982, 1985, 1988a, 1989, 1990, 1991). The terms competitive and cooperative system seem to have been around before this in the population biology literature.

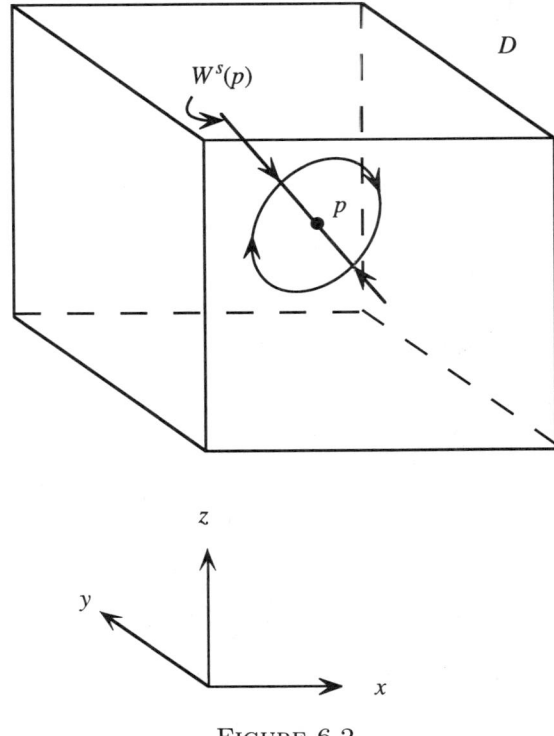

FIGURE 6.2

Proposition 2.1 is an amalgamation of two results which the author only recently realized were related to each other. The older result is the part of the Proposition concerned with competitive systems and is due to Grossberg(1978,1980). Grossberg calls the i-th component of x "on" if $f_i(x) > 0$ and "off" if $f_i(x) < 0$. The proposition asserts that a solution of a competitive system which doesn't converge monotonically to an equilibrium must always have at least one component on and at least one component off. The part of the proposition dealing with cooperative systems is due to Selgrade(1980). The papers Selgrade(1980,1982,1983) represent pioneering work on monotone dynamics for ordinary differential equations. Theorem 2.2 for two-dimensional systems has been proved many times. See Albrecht et al(1974), Kolmogorov(1936), Selgrade(1980), Hirsch and Smale(1974) and Hirsch(1982). The proof here seems to be new.

The results of section three are due to Hirsch(1982). Hirsch attributes the proof of Lemma 3.1 to L.Ito. The proof given, a slight variant of Ito's proof, appears in Smith and Waltman(1995).

A remarkable result for competitive systems in $\mathbb{R}^n_+$ of the form $x'_i = x_i N_i(x)$ is obtained in Hirsch(1988a). He assumes that the community matrix $(\frac{\partial N_i}{\partial x_j}(x))$ has nonpositive off-diagonal elements and is irreducible (see Chapter 4) at each $x \geq 0$. The system is also assumed to be dissipative. In that case it is shown that there exist a countable family of open invariant Lipschitz $(n-1)$-cells such that every persistent non-convergent trajectory is asymptotic to a trajectory in one of the $(n-1)$-cells. A persistent trajectory is one whose omega limit set lies in the interior of $\mathbb{R}^n_+$. Furthermore, each cell is unordered and the cardinality of the family of cells is at most one plus the number of weak sources in the interior of $\mathbb{R}^n_+$.

Theorem 4.1 is due to Hirsch(1982, 1990). Following the partial result in the earlier paper of Hirsch, the conclusion of Theorem 4.1 was proved in Smith(1986e) under the additional assumption of irreducibility. This assumption was removed in Hirsch(1990) and our proof follows the one given there.

Theorem 4.2 is a slightly more general result than that given in Smith and Waltman(1995). The positively invariant set K was first described in Smith(1986e). Using degree theory, it was shown there that if K contains a unique equilibrium which is hyperbolic, then its stable manifold has dimension one or three. See Smith and Waltman(1987) for an application to three dimensional Kolmogorov competitive systems.

Smith(1986d) used non-standard cones in monotone systems theory to analyze a model of n interacting populations, k of which interact with each other in a cooperative manner, the remaining $n-k$ interact with each other in a cooperative manner, but the interactions between any two populations in different groups is competitive. The review article Smith(1988) develops these ideas further where the algorithm (5.3) is described. The graph-theoretic test seems to be due to Travis and Post(1979).

The treatment of the Field-Noyes model follows that in the thesis of Zhu(1991). It can be shown that either p is stable or there exists an orbitally asymptotically stable periodic solution. Moreover, there can exist only finitely many periodic orbits. These results are an application of the main results of Zhu and Smith(1994). Additional applications of Theorem 4.1 appear in Hsu and Waltman(1992), Smith and Waltman(1995) and Li and Muldowney(1994).

Zeeman(1993) considers the three species Lotka-Volterra competitive system, identifying 33 stable equivalence classes (within the space of Lotka-Volterra competitive systems) of dynamical behavior based on the geometry of the isoclines.

CHAPTER 4

Irreducible Cooperative Systems

0. Introduction

We continue our study of cooperative and competitive systems in this chapter, with special emphasis on the former. Our first task is to give sufficient conditions for a cooperative system of differential equations to generate a strongly monotone flow. For this to hold, the Jacobian matrix of the vector field must be irreducible at each point. The principal result of Chapter 2 can then be applied to show that the generic orbit of a cooperative and irreducible system converges to equilibrium.

The irreducibility hypothesis also plays an important role in stability considerations for cooperative and irreducible systems. The stability of an equilibrium of such a system is determined by the dominant eigenvalue of the Jacobian matrix. The Perron-Frobenius theory asserts that this dominant eigenvalue is real and simple and there is a corresponding positive eigenvector. These facts lead to information on the basin of attraction of the equilibrium in the case that it is attracting and to the existence of monotone heteroclinic orbits belonging to the unstable manifold of the equilibrium in the case that the latter is unstable. The Perron-Frobenius Theorem also implies that a periodic orbit of a cooperative and irreducible system has a positive Floquet multiplier larger than one. In other words, it is unstable in the linear approximation, amplifying the result of Chapter 1 that a monotone dynamical system cannot have an attracting periodic orbit.

One of the earliest results in the theory of competitive systems of differential equations was due to S. Smale. In a cautionary note to population biologists, he showed that essentially arbitrary dynamics, including chaos, is possible in models of n competing populations, for n sufficiently large, if one merely assumes that an increase in population x_j decreases the rate of change of population x_i and that the system is dissipative. This construction of Smale, presented in section 5, implies that the main results of Chapters 1 and 2 cannot be improved to conclude that all solutions of a strongly monotone dynamical system converge to equilibrium without substantial further assumptions. Chaos is possible in cooperative and irreducible systems but, if it exists, it is so unstable as to be unobservable. Competitive systems, on the other hand, can have strange attractors.

The results of this chapter are applied to a biochemical control loop in section 2. This is a model of the control of protein synthesis in the cell. It is shown that the generic solution converges to an equilibrium. In section 4, a Lotka-Volterra model of two competing populations occupying a two-patch environment is considered. Conditions are given for the generic solution of this four dimensional system to approach a positive equilibrium representing the coexistence of both populations in each patch.

1. Strong Monotonicity

We call the system
$$x' = f(x) \tag{1.1}$$
irreducible in D provided the Jacobian matrix $Df(x)$ is an irreducible matrix for every $x \in D$. Recall that an $n \times n$ matrix $A = (a_{ij})$ is irreducible if for every nonempty, proper subset I of the set $N = \{1, 2, \ldots, n\}$, there is an $i \in I$ and $j \in J \equiv N \setminus I$ such that $a_{ij} \neq 0$. We write $A \geq 0$ if each entry of A is non-negative, $A > 0$ if $A \geq 0$ and A has at least one non-zero entry, and $A \gg 0$ if each entry of A is positive.

If (1.1) is cooperative in D, that is if (3.1.3) is satisfied, then it follows from Remark 3.1.1 that
$$\frac{\partial \phi_t}{\partial x}(x) > 0, \quad t \geq 0. \tag{1.2}$$
In fact, the Jacobian matrix of ϕ_t is the fundamental matrix solution of the linear variational equation
$$z' = Df(\phi_t(x))z. \tag{1.3}$$
Therefore, its i-th column vector, $q_i(t)$ satisfies (1.3) and $q_i(0) = e_i$, the i-th standard basis vector. Since $q_i(0) > 0$, Remark 3.1.3 implies that $q_i(t) > 0$ for $t > 0$, which leads to (1.2). The Jacobian matrix of ϕ_t is strongly positive if (1.1) is cooperative and irreducible.

THEOREM 1.1. *Let (1.1) be cooperative and irreducible in D. Then*
$$\frac{\partial \phi_t}{\partial x}(x) \gg 0, \quad t > 0. \tag{1.4}$$
Furthermore, if $x_0, y_0 \in D$ satisfy $x_0 < y_0$, $t > 0$ and if $\phi_t(x_0), \phi_t(y_0)$ are defined, then
$$\phi_t(x_0) \ll \phi_t(y_0), \quad t > 0.$$

PROOF. The second assertion follows from (1.4) and the formula
$$\phi_t(y_0) - \phi_t(x_0) = \int_0^1 \frac{\partial \phi_t}{\partial x}(x_0 + r(y_0 - x_0))(y_0 - x_0))dr.$$
Let $X(t) = \frac{\partial \phi_t}{\partial x}(x)$ and $A(t) = Df(\phi_t(x))$. Then, $X(t) = (x_{ij}(t))$ satisfies
$$X'(t) = A(t)X(t), \quad X(0) = I.$$
By (3.1.3) and (1.2),
$$x'_{ij}(t) \geq a_{ii}(t)x_{ij}(t), \quad t \geq 0.$$
If $t_1 \geq 0$ and $x_{ij}(t_1) > 0$, then it follows that $x_{ij}(t) > 0$ for $t \geq t_1$. Consequently, $Z_{ij} \equiv \{t : t > 0 \text{ and } x_{ij}(t) = 0\}$ is either empty or is an interval $(0, e_{ij}]$, $0 < e_{ij} \leq \infty$. Obviously, Z_{ii} is empty since $x_{ii}(0) = 1$, $1 \leq i \leq n$. Suppose that Z_{ij} is not empty for some pair of indices i, j. Let $S = \{k : x_{kj}(t) = 0 \text{ for all } t \in (0, e_{ij}]\}$. As $i \in S$, S is not empty and as $X(t)$ is nonsingular, $S \neq \{1, 2, \ldots, n\}$. Let S^c denote the complement of S in $\{1, 2, \ldots, n\}$. For $t \in (0, e_{ij}]$, $k \in S$,
$$0 = x'_{kj}(t) = \sum_l a_{kl}(t)x_{lj}(t) = \sum_{l \in S^c} a_{kl}(t)x_{lj}(t).$$
But, by the definition of S, it follows that $x_{lj}(t) > 0$ for t near e_{ij}, $l \in S^c$. Consequently, there exists $t_1 < e_{ij}$ such that $x_{lj}(t) > 0$ for $t_1 < t$ and all $l \in S^c$. It

follows from the equality above that $a_{kl}(t) = 0$ for $t_1 < t \leq e_{ij}$, and for all $l \in S^c$. As $k \in S$ was arbitrary, we have that $a_{kl}(t) = 0$ for $t_1 < t \leq e_{ij}$ for all $k \in S$ and all $l \in S^c$. This contradicts the irreducibility of $A(t)$ and completes the proof. □

Note from the proof of Theorem 1.1 that (1.4) holds if for every interval (a,b) there exists $t_0 \in (a,b)$ such that $A(t_0)$ is irreducible.

According to Theorem 1.1, the flow of a cooperative and irreducible system is strongly monotone. In order to apply the results of Chapters 1 and 2, sufficient conditions for the compactness requirement (C) of Chapter 1 are required. The following mild condition will do the job.

(T) For each $x \in D$, $\phi_t(x)$ is defined for all $t \geq 0$ and $\phi_t(x) \in D$. Moreover, for each bounded subset A of D, there exists a closed and bounded subset $B = B(A)$ of D such that for each $x \in A$, $\phi_t(x) \in B$ for all large t.

The main result of the Chapter follows. In it, we assume that either the domain D is an open subset of $\mathbb{R}^n$ or it is the closure of an open subset of $\mathbb{R}^n$. In the latter case, it is assumed that the vector field extends to a continuously differentiable one on a neighborhood of D. Recall that D is necessarily p-convex if (1.1) is cooperative (see Remark 1.1 of Chapter 3).

THEOREM 1.2. *Let* (1.1) *be cooperative and irreducible on D and assume that* (T) *holds. Assume also that each point $x \in D$ can be approximated either from above or from below by a sequence $\{x_n\}$ in D. Then the set C of convergent points contains an open and dense subset of D.*

PROOF. Once we show that (C) holds, the result will follow from Theorem 2.4.7 and Theorem 1.1. The latter implies that the semiflow is strongly monotone and (1.4) implies that the Jacobian is strongly positive.

To show that (C) holds first observe that (T) implies that all forward orbits have compact closure in D. This follows by taking $A = \{x\}$ for $x \in D$. If $\{x_n\}_{n \geq 1}$ is a sequence approximating $x_0 \in D$ from either above or from below in D, then let $A = \{x_n\}_{n \geq 0}$. As A is compact in D, there exists a closed and bounded subset B of D such that for each $x \in A$, $\phi_t(x) \in B$ for all large t. Let B' be the closure in D of an ϵ-neighborhood of B. Our assumptions concerning D ensure that B' is a compact subset of D. By the continuity of ϕ_t, it follows that there exists $t_0 > 0$ such that $\phi_t(x) \in B'$ for all $t \geq t_0$ and for all $x \in A$. Consequently, $\cup_{n \geq 0} \omega(x_n) \subset B'$ and (C) follows immediately. □

Theorems 1.1 and 1.2 have obvious analogs in case that (1.1) is cooperative with respect to K_m for some m (see Chapter 3, section 5) and irreducible. We will use these analogous results later in the Chapter.

A common difficulty in applying Theorem 1.2, often overlooked in applications, arises from the fact that irreducibility is an open condition. It commonly occurs that irreducibility holds only in the interior of the domain while the boundary is an invariant set. In this case, strong monotonicity (and even the strong order preserving property) may fail to hold on the boundary. A simple example will illustrate the problem. Consider the following model of the mutualistic interaction of n populations x_i, $1 \leq i \leq n$, given by

$$x'_i = x_i \left(r_i + \sum_{j=1}^n a_{ij} x_j \right) \quad (1.5)$$

where $r_i \in \mathbb{R}$, $a_{ij} \geq 0$ for $i \neq j$ and the matrix $A = (a_{ij})$ is irreducible. The domain of interest is $D = \mathbb{R}^n_+$, the boundary of which is invariant for (1.5). The system is obviously cooperative in D. If $f(x)$ denotes the vector field, then $Df(x)$ is irreducible in the interior of D but not on the boundary of D. In fact, the flow is not strongly monotone nor strongly order preserving in D. It is easier to see this in the simple case $n = 2$. Consider the solutions with initial data given by $x = (x_1, 0)$ and $\hat{x} = (\hat{x}_1, 0)$ where $0 < x_1 < \hat{x}_1$. Then $x < \hat{x}$ and obviously $\phi_t(x) \ll \phi_t(\hat{x})$ can never hold since the x_1-axis is invariant. Thus ϕ is not strongly monotone. If ϕ were SOP in D then there would exist neighborhoods U of x and V of $\hat{x}$ in D and $t_0 \geq 0$ such that $\phi_{t_0}(U) \leq \phi_{t_0}(V)$. But $\phi_{t_0}(U)$ contains a point p satisfying $p \gg 0$ so the inequality $p \leq \phi_{t_0}(\hat{x})$, required for SOP to hold, cannot hold.

The flow is strongly monotone in the positively invariant set $\text{Int}\mathbb{R}^n_+$ but unfortunately, the omega limit set of a point of this set may belong to the boundary or may intersect the boundary. An examination of the proofs in Chapter 1 suggests that the strong order preserving property must hold on the limit set for these results to hold. The point is that Theorem 1.2 cannot be immediately applied to (1.5) because the irreducibility hypothesis does not hold on all of D. See section 6 for further remarks concerning (1.5).

Although strong monotonicity fails to hold for (1.5) on D, a weaker condition holds which is often useful. We state it in some generality.

REMARK 1.1. Suppose that (1.1) is cooperative on the positively invariant set D, where D is the closure of an open set G on which f is irreducible. If $x, y \in D$, $x < y$ and at least one of the points belongs to G, then $\phi_t(x) \ll \phi_t(y)$ for $t > 0$. This follows immediately from the proof of Theorem 1.1.

2. A Biochemical Control Circuit

As an application of Theorem 1.2, we consider a mathematical model of the control of protein synthesis in the cell. The biology is as follows. A segment of DNA is assumed to be translated to mRNA which in turn is translated to produce an enzyme and it in turn is translated to another enzyme and so on until an end product molecule is produced. This end product acts on a nearby segment of DNA to produce a feedback loop, controlling the translation of DNA to mRNA. In the case of interest here, the feedback is assumed to be positive. Let x_1 be the cellular concentration of mRNA, x_2 the concentration of the first enzyme, and so on, finally x_n is the concentration of end product. Then the biochemical control circuit is described by the system of equations

$$\begin{aligned} x'_1 &= g(x_n) - \alpha_1 x_1 \\ x'_i &= x_{i-1} - \alpha_i x_i, \quad 2 \leq i \leq n, \end{aligned} \quad (2.1)$$

where $\alpha_i > 0$ and g is a bounded continuously differentiable function satisfying

$$M > g(u) > 0, \quad g'(u) > 0, \quad u > 0.$$

2. A BIOCHEMICAL CONTROL CIRCUIT

The nonlinearity g provides the feedback of the end product on the translation of mRNA. The translation becomes more efficient as the concentration of end product increases. Typical choices of g in the applications are

$$g(u) = \frac{u^p}{1+u^p}, \tag{2.2}$$

or

$$g(u) = \frac{1+u^p}{K+u^p}, \tag{2.3}$$

where p is a positive integer and $K > 1$. The x_i represent concentrations of various macro-molecules in the cell and therefore must be nonnegative. It is clear from the form of (2.1) that $\mathbb{R}^n_+$ is positively invariant. Furthermore, (2.1) is cooperative and irreducible in $\mathbb{R}^n_+$ and, as a consequence of Theorem 1.1, the flow ϕ_t is strongly monotone there.

Equilibria of (2.1) are in one-to-one correspondence with solutions of

$$g(u) = \alpha u \tag{2.4}$$

where $\alpha = \prod \alpha_i$. For each solution of (2.4) there is an equilibrium given by

$$x_n = u, x_{n-1} = \alpha_n x_n, \ldots, x_1 = \alpha_2 x_2.$$

From this observation, the set E of equilibria is totally ordered under the ordering $<$ and is nonempty. Indeed, either $g(0) = 0$ in which case 0 is an equilibrium, or $g(0) > 0$ in which case the smallest equilibrium is positive in each coordinate. Let E_* and E_{**} denote the smallest and largest elements of E. The latter exists since g is bounded. Then $0 \leq E_* \leq E_{**}$, where equality may hold in each inequality. Clearly,

$$E \subset [E_*, E_{**}].$$

There is a one parameter family of positively invariant order intervals for (2.1) given by

$$B(r) = r[0, w], \quad r \geq 1,$$

where

$$w = M(\alpha_1^{-1}, (\alpha_1\alpha_2)^{-1}, \ldots, (\alpha_1\alpha_2\ldots\alpha_n)^{-1}).$$

In fact, if $f(x)$ denotes the vector field on the right side of (2.1), then an easy calculation shows that

$$f(rw) < 0 \leq f(0), \quad r \geq 1.$$

In fact, this same calculation shows that $E_{**} < w$. By Proposition 3.2.1, $\phi_t(0)$ converges monotonically to the smallest equilibrium E_* and $\phi_t(rw)$ converges monotonically to the largest equilibrium E_{**} ($E_{**} < rw$, $r \geq 1$).

If $x \in \mathbb{R}^n_+$, then there exists $r \geq 1$ such that $0 < x < rw$ for some $r \geq 1$. By monotonicity,

$$0 < \phi_t(0) < \phi_t(x) < \phi_t(rw) < rw, \quad t > 0.$$

Letting $t \to \infty$ leads to

$$\omega(x) \subset [E_*, E_{**}]$$

for each $x \in \mathbb{R}^n_+$. This implies that the system (2.1) is *dissipative*, that is, there is a bounded set B such that $\phi_t(x) \in B$ for all large t, and for each $x \in \mathbb{R}^n_+$. As a consequence, the compactness hypotheses of Theorem 1.2 hold and we have the following result.

PROPOSITION 2.1. *All orbits of* (2.1) *are attracted to* $[E_*, E_{**}]$. *There is an open and dense subset of* $\mathbb{R}_+^n$ *consisting of convergent points for* (2.1). *If E is a single point, then it attracts all solutions. If E consists of two points, then all solutions are attracted to one of these points.*

PROOF. The first assertion follows from Theorem 1.2. The second assertion follows from Theorem 2.3.1, the third from Theorem 2.3.2. □

3. Stability and the Perron-Frobenius Theorem

The Perron-Frobenius Theorem plays an important role in the stability theory for competitive and cooperative systems. Before stating it, some notation will be useful. Let A be an $n \times n$ matrix. The set of eigenvalues of A will be denoted by $\sigma(A)$. The spectral radius of A is defined by

$$\rho(A) = \max\{|\lambda| : \lambda \in \sigma(A)\}.$$

THEOREM 3.1. (*Perron-Frobenius*) *If A is an* $n \times n$ *nonnegative matrix, then* $r = \rho(A)$ *is an eigenvalue of A and there is a corresponding eigenvector* $v > 0$. *If, in addition, A is irreducible then* $r > 0$ *and* $v \gg 0$. *Moreover, r has algebraic multiplicity one and if* $u > 0$ *is an eigenvector of A, then there exists* $s > 0$ *such that* $u = sv$. *If B is a matrix satisfying* $B > A$, *then* $\rho(B) > \rho(A)$. *Finally, if* $A \gg 0$ *then* $|\lambda| < r$ *for all other eigenvalues of A.*

See Berman and Plemmons(1979) for a proof. Note that this result is the finite dimensional version of the Krein-Rutman Theorem (Theorem 2.4.1). Obviously, if $A \gg 0$ then A is irreducible.

If x_0 is an equilibrium of a cooperative system (1.1), then the Jacobian matrix $A = Df(x_0)$ has nonnegative off-diagonal entries. Hereafter, we call such a matrix *quasi-positive*. The *stability modulus* of A is defined by

$$s(A) = \max\{\Re\lambda : \lambda \in \sigma(A)\}.$$

Here, $\Re\lambda$ denotes the real part of λ. Clearly, $s(A)$ determines the stability properties of the equilibrium x_0. The next result is a consequence of the Perron-Frobenius Theorem.

COROLLARY 3.2. *Let A be a quasi-positive matrix. Then* $s(A) \in \sigma(A)$ *and there is a vector* $v > 0$ *such that*

$$Av = s(A)v.$$

Moreover, $\Re\lambda < s(A)$ *for all* $\lambda \in \sigma(A) \setminus \{s(A)\}$. *If, in addition, A is irreducible, then*
(1) $s(A)$ *has algebraic multiplicity one.*
(2) $v \gg 0$ *and any eigenvector* $w > 0$ *of A is a positive multiple of v.*
(3) *If B is a matrix satisfying* $B > A$, *then* $s(B) > s(A)$.
(4) *If* $s(A) < 0$, *then* $-A^{-1} \gg 0$.

PROOF. $A + cI \geq 0$ for all large c since A is quasi-positive. Therefore, Theorem 3.1 applies to $A + cI$ for such c. In particular, the spectral radius is positive and it is an eigenvalue of $A + cI$ for all large c. Moreover, there exists a corresponding nonnegative eigenvector. Since adding cI to A results in $\sigma(A + cI) = c + \sigma(A)$, it follows that $s(A + cI) = \rho(A + cI) = s(A) + c$ and therefore, $s(A)$ is an eigenvalue of

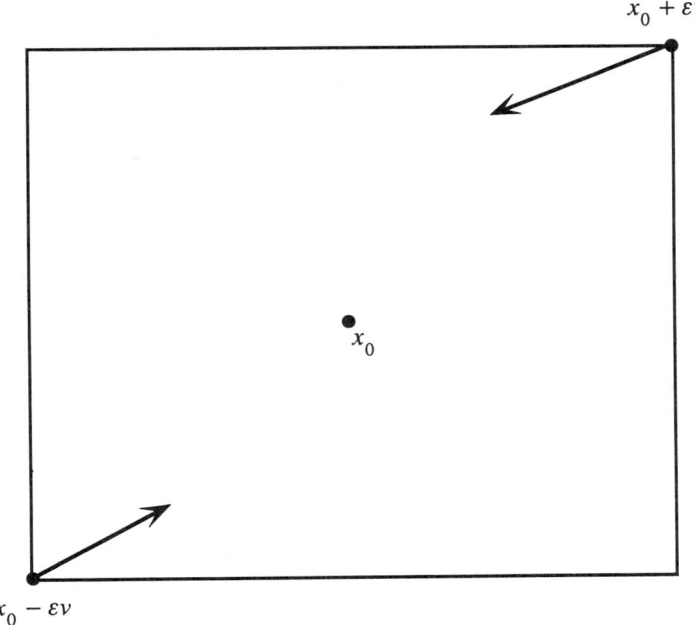

FIGURE 3.1. The vectors represent $f(x_0 \pm \epsilon v)$.

A. Also, adding cI to A preserves the property of irreducibility in the sense that A is irreducible if and only if $A + cI$ is irreducible. These observations and Theorem 3.1 lead immediately to (1),(2) and (3).

(4) follows from (1.4) and the formula $-A^{-1} = \int_0^\infty e^{At} dt$, which is easily checked by multiplying by A and using the fundamental theorem of calculus. □

According to Corollary 3.2, the stability of an equilibrium x_0 of a cooperative system is governed by a real eigenvalue of the variational matrix $Df(x_0)$. In the remaining results of this section it is crucial that there is a positive eigenvector corresponding to $s(Df(x_0))$. A sufficient, but not necessary, condition for this to hold is that $Df(x_0)$ is irreducible.

REMARK 3.1. If x_0 is an equilibrium of (1.1), $s = s(Df(x_0)) < 0$ and there is an eigenvector $v \gg 0$ such that $Df(x_0)v = sv$, then there exists $\epsilon > 0$ such that the order interval $[x_0 - \epsilon v, x_0 + \epsilon v]$ is a positively invariant subset of the basin of attraction of x_0. A calculation yields that

$$f(x_0 + rv) = rDf(x_0)v + o(r) = r(sv + O(r))$$

where $o(r)/r \to 0$ and $O(r) \to 0$ as $r \to 0$. Consequently, there is an $\epsilon > 0$ such that $f(x_0 + rv) \gg 0$ for $r \in [-\epsilon, 0)$ and $f(x_0 + rv) \ll 0$ for $r \in (0, \epsilon]$. Theorem 3.2.1 implies that $\phi_t(x_0 + \epsilon v)$ monotonically decreases and $\phi_t(x_0 - \epsilon v)$ monotonically increases to x_0. Therefore the order interval is a positively invariant set. If $x_0 - \epsilon v \leq x \leq x_0 + \epsilon v$, then $\phi_t(x_0 - \epsilon v) \leq \phi_t(x) \leq \phi_t(x_0 + \epsilon v)$ for $t > 0$ and therefore $\phi_t(x) \to x_0$ as $t \to \infty$. This proves the assertion which is illustrated in **Figure 3.1**.

If $s(Df(x_0)) > 0$, then x_0 is unstable. If there is a corresponding positive eigenvector, then the next result asserts the existence of monotonically converging solutions issuing from points on the ray though x_0 in the direction of the eigenvector. It is very useful in applications.

THEOREM 3.3. *Let (1.1) be cooperative in D and let $x_0 \in D$ be an equilibrium. Suppose that $s = s(Df(x_0)) > 0$ and there is an eigenvector $v \gg 0$ such that $Df(x_0)v = sv$. Also, assume that for some $\epsilon > 0$, $\gamma^+(x_r)$ has compact closure in D for each $x_r \equiv x_0 + rv$, $r \in (0, \epsilon]$. Then there exists $\epsilon_0 \in (0, \epsilon]$ and $e \in E$ such that for each $r \in (0, \epsilon_0]$, the solution $\phi_t(x_r)$ has the following properties:*
(1) $x_r \ll \phi_t(x_r) \ll \phi_s(x_r) \ll e$, $0 < t < s$.
(2) $\frac{d}{dt}\phi_t(x_r) \gg 0$, $t > 0$.
(3) $\phi_t(x_r) \to e$, $t \to \infty$.
If, in addition, $Df(x_0)$ is irreducible, then there exists y satisfying $x_0 \ll y \ll e$ such that $\frac{d}{dt}\phi_t(y) \gg 0$ for all $t \in \mathbb{R}$, $\phi_t(y) \to e$ as $t \to \infty$ and
(4) $\phi_t(y) \to x_0$ *as $t \to -\infty$. Furthermore, $\phi_t(y)$ approaches x_0 tangent to the eigenvector v.*

PROOF. A calculation yields
$$f(x_r) = Df(x_0)rv + o(r)$$
$$= srv + o(r)$$
$$= r(sv + O(r)) \gg 0$$
if $0 < r \le \epsilon_0$ for suitable $\epsilon_0 \le \epsilon$. Here, $o(r)$ represents a term satisfying $o(r)/r \to 0$ as $r \to 0$, and $O(r)$ represents a term satisfying $O(r) \to 0$ as $r \to 0$. According to Proposition 3.2.1,
$$\frac{d}{dt}\phi_t(x_r) = f(\phi_t(x_r)) \gg 0$$
for all $t > 0$. Therefore there exists $e_r \in E$ such that (1),(2) and (3) hold with $e = e_r$, for each $r \in (0, \epsilon_0]$. To show that e_r is independent of r, first note that $0 < r_1 < r_2 \le \epsilon_0$ implies that $x_{r_1} \ll x_{r_2}$ and $\phi_t(x_{r_1}) \ll \phi_t(x_{r_2})$ so, on letting $t \to \infty$, $e_{r_1} \le e_{r_2}$. Since $x_r = x_0 + rv \ll e_r$, we can choose $h > 0$ such that $r + h \le \epsilon_0$ and $x_{r+h} \le e_r$ and such that h is maximal with these properties. Then, $e_r = \phi_t(e_r) \ge \phi_t(x_{r+h})$ for $t \ge 0$ so letting $t \to \infty$ leads to $e_r \ge e_{r+h}$. But $e_{r+h} \ge e_r$, as noted above, so we conclude that $e_{r+h} = e_r$. Consequently, $e_r = e_{r+h} \gg x_{r+h}$ and therefore, the maximality of h implies that $r + h = \epsilon_0$. Thus $e_r = e_{\epsilon_0} \equiv e$ for $r \in (0, \epsilon_0]$.

Now, suppose that $Df(x_0)$ is irreducible. Then, by Corollary 3.2 (1), s has algebraic multiplicity one and the corresponding eigenspace is spanned by v. By Coddington and Levinson(1955), Theorem 4.4, Chapter 13, applied to the time-reversed flow, there is a one dimensional, continuously differentiable manifold, S, containing x_0, tangent at x_0 to the ray $x = x_r$, $r \ge 0$, each point of which satisfies
$$\liminf_{t \to -\infty} t^{-1} \log |\phi_t(x) - x_0| = s.$$

Actually, for x different from x_0 on S the liminf can be replaced by lim (see Coppel(1965) p.97). Since S is tangent at x_0 to the line through x_0 in the direction v, the point x_0 separates S into two parts, S_+ and S_- where $x \gg 0$ ($x \ll 0$) for all $x \in S_+$ ($x \in S_-$). Points x of S_+ sufficiently near x_0 satisfy $f(x) \gg 0$ by an

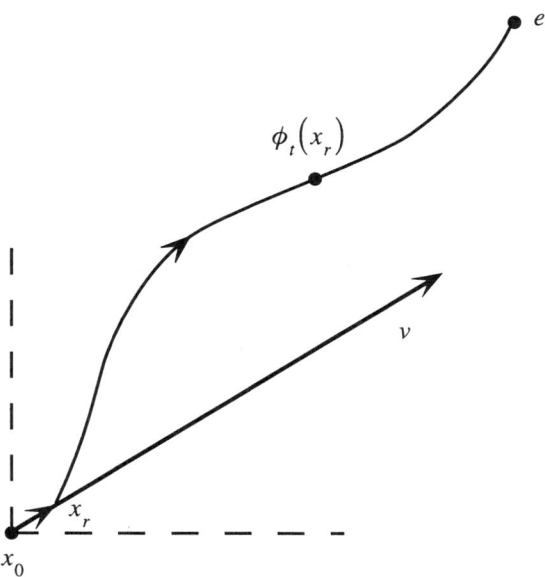

FIGURE 3.2. A monotone orbit starting at $x_r = x_0 + rv$ and converging to e.

argument similar to the one used in the first calculation of the proof since S is tangent to the ray x_r. It follows that if we let y be any such point, then (4) holds and $f(\phi_t(y)) \gg 0$ for all t, by Proposition 3.2.1. By choosing r sufficiently small we can ensure that $x_r \ll y \ll e$. Monotonicity implies that $\phi_t(x_r) \ll \phi_t(y) \ll e$ for $t \geq 0$ and since $\phi_t(x_r) \to e$ as $t \to \infty$, it follows that $\phi_t(y) \to e$ as $t \to \infty$. $\square$

Figure 3.2 illustrates the conclusions of the Theorem.

REMARK 3.2. An analogous result holds if we take the eigenvector $u = -v \ll 0$. We must assume that for some $\epsilon > 0$, $\gamma^+(x_r)$ has compact closure in D for each $x_r \equiv x_0 + ru$, $r \in (0, \epsilon]$. Then, there exists ϵ_0 and $e \in E$, as in Theorem 3.3, such that (1),(2) and (3) hold with inequalities reversed.

REMARK 3.3. The equilibrium e in Theorem 3.3 attracts all solutions corresponding to initial conditions $x \in D$ satisfying $x_0 \ll x \leq e$. This is a consequence of the existence of r such that $x_r \ll x \leq e$ and a comparison argument. In particular, $s(Df(e)) \leq 0$ by Remark 3.2.

REMARK 3.4. Theorem 3.3 and its proof should be compared with the abstract result Theorem 2.5.3. Observe that hypothesis (U_+) of the latter result is satisfied with $\eta(s) = x_0 + (s\epsilon_0)v$ where ϵ_0 is as in the proof of Theorem 3.3 since the points x_r are strict sub-equilibria. Note however that we do not assume that (1.1) is irreducible in $\{x \in D : x_0 \ll x\}$ in Theorem 3.3, as would be required to apply Theorem 2.5.3.

Theorem 1.2.2 implies that a cooperative system cannot have an attracting periodic orbit. In fact, if the system is irreducible, then a periodic solution is unstable in the linear approximation as the next result shows.

PROPOSITION 3.4. *A periodic solution of a cooperative and irreducible system is unstable in the linear approximation. More precisely, it has a Floquet multiplier, ρ, of algebraic multiplicity one that is strictly larger in modulus than all other multipliers and satisfies $\rho > 1$.*

PROOF. Let $\phi_t(p)$ be a periodic solution and let $T > 0$ be the minimal period of $\phi_t(p)$. By Theorem 1.1, $\Phi = \frac{\partial \phi_T}{\partial x}(p) \gg 0$. The Floquet multipliers are the eigenvalues of Φ. By Theorem 3.1, the matrix Φ has a positive eigenvalue, $\rho > 0$ which has algebraic multiplicity one and is larger than the modulus of all other eigenvalues. Furthermore, the corresponding eigenspace is spanned by a vector $v \gg 0$. Now, $\Phi f(p) = f(p)$ since $\frac{d}{dt}\phi_t(p)$ is a solution of (1.3) with $x = p$ and $f(p) \notin \mathbb{R}^n_+ \cup (-\mathbb{R}^n_+)$ by Proposition 3.2.1. It follows that $\rho > 1$. □

By reversing the direction of time, Proposition 3.4 implies that a periodic orbit of a competitive and irreducible system always has a positive Floquet multiplier of algebraic multiplicity one that is less than one and which has the smallest modulus among all other multipliers.

As noted in the previous section, all the results of the present section have obvious analogs in the case that (1.1) is cooperative with respect to K_m. We will use these results in the next section.

4. Competition and Migration

Consider two competing populations occupying an environment consisting of two discrete patches between which they can migrate. Let x_i, $i = 1, 2$, denote the population density of one population in patches 1 and 2 respectively and y_i, $i = 1, 2$, denote the population density of the second population in each of the two patches. If no mortality is suffered during migration between patches and if Lotka-Volterra dynamics is used to describe competition, then the following equations describe the time evolution of the population densities.

$$\begin{aligned} x_1' &= \epsilon(x_2 - x_1) + r_1 x_1 (1 - x_1 K_1^{-1} - a_1 y_1) \\ x_2' &= \epsilon(x_1 - x_2) + r_2 x_2 (1 - x_2 K_2^{-1} - a_2 y_2) \\ y_1' &= \delta(y_2 - y_1) + s_1 y_1 (1 - y_1 L_1^{-1} - b_1 x_1) \\ y_2' &= \delta(y_1 - y_2) + s_2 y_2 (1 - y_2 L_2^{-1} - b_2 x_2) \end{aligned} \quad (4.1)$$

The Jacobian matrix of the right side of (4.1) has the form:

$$\begin{pmatrix} * & \epsilon & -r_1 a_1 x_1 & 0 \\ \epsilon & * & 0 & -r_2 a_2 x_2 \\ -s_1 b_1 y_1 & 0 & * & \delta \\ 0 & -s_2 b_2 y_2 & \delta & * \end{pmatrix}$$

where "*" represents terms on the main diagonal. It is apparent that (4.1) is cooperative with respect to K_m where K_m is the cone

$$K_m = \{z = (x, y) \in \mathbb{R}^4_+ : x \geq 0, \ y \leq 0\}.$$

Recall, we write $z = (x, y) \leq_m \bar{z} = (\bar{x}, \bar{y})$ if $x \leq \bar{x}$ and $y \geq \bar{y}$. Furthermore, the Jacobian matrix is irreducible in Int$\mathbb{R}^4_+$. Therefore, the flow $\phi_t(z)$ of (4.1) in $\mathbb{R}^4_+$ is monotone with respect to the order relation $\leq_m$ and is strongly monotone in the interior of $\mathbb{R}^4_+$ by Theorem 1.1 and Proposition 3.5.1.

Consider the equilibria of (4.1). If $z = (x, y)$ is an equilibrium and if $x_1 > 0$ ($x_2 > 0$), then necessarily $x_2 > 0$ ($x_1 > 0$) by the form of (4.1). Therefore, nontrivial equilibria of (4.1) have the form $(\hat{x}_1, \hat{x}_2, 0, 0)$, $(0, 0, \tilde{y}_1, \tilde{y}_2)$, and $(x_1^*, x_2^*, y_1^*, y_2^*)$ where each component not explicitly zero must be positive.

Before treating the full system (4.1), it is useful to consider the subsystem obtained by setting $y = 0$ in (4.1). This planar system is given by

$$\begin{aligned} x_1' &= \epsilon(x_2 - x_1) + r_1 x_1 (1 - x_1 K_1^{-1}) \\ x_2' &= \epsilon(x_1 - x_2) + r_2 x_2 (1 - x_2 K_2^{-1}) \end{aligned} \quad (4.2)$$

Obviously, (4.2) is cooperative in $\mathbb{R}_+^2$ and cooperative and irreducible in $\text{Int}\,\mathbb{R}_+^2$. The next result describes the behavior of solutions of (4.2).

PROPOSITION 4.1. *There is a unique non-zero equilibrium $\hat{x}$. $\hat{x} \gg 0$ and it attracts all non-trivial solutions of (4.2).*

PROOF. The form of (4.2) guarantees that $\mathbb{R}_+^2$ is positively invariant. Let f_1 denote the vector field described by (4.2), ϕ_t the corresponding flow, and let $\mathbf{1} = (1, 1)$. Then for all large positive m, $f_1(m\mathbf{1}) \ll 0$. Consequently, Proposition 3.2.1 implies that $\phi_t(m\mathbf{1})$ converges monotonically to an equilibrium for all such m. By a comparison argument it follows that (4.2) is dissipative, all positive orbits are bounded. The Jacobian matrix of f_1 at 0 is given by

$$J = \begin{pmatrix} r_1 - \epsilon & \epsilon \\ \epsilon & r_2 - \epsilon \end{pmatrix}.$$

It is easy to see that $s(J) > 0$ since whenever the determinant of J is nonnegative, then the trace of J is positive. As J is irreducible, Corollary 3.2 implies the existence of a positive eigenvector v corresponding to $s(J)$ and therefore Theorem 3.3 implies the existence of at least one positive equilibrium. In fact, it is easy to see directly that there is a positive equilibrium but it is difficult to give an explicit formula for one. Direct calculation shows that there is an equilibrium $\hat{x}$ given by

$$\begin{aligned} K_1^{-1} \hat{x}_1 &= 1 + \frac{\epsilon}{r_1}(u^{-1} - 1) \\ K_2^{-1} \hat{x}_2 &= 1 + \frac{\epsilon}{r_2}(u - 1) \end{aligned} \quad (4.3)$$

where $u = x_1/x_2$ and u satisfies

$$p(u) = u^3 + \left(\frac{r_2}{\epsilon} - 1\right)u^2 - \left(\frac{r_1}{\epsilon} - 1\right)Q^{-1}u - Q^{-1} = 0. \quad (4.4)$$

Q is the quotient $\frac{r_1 K_2}{r_2 K_1}$. Conversely, every root of (4.4) for which the right sides of (4.3) are positive corresponds to an equilibrium. If u_i, $i = 1, 2, 3$, denote the roots of (4.4), then

$$\begin{aligned} u_1 u_2 u_3 &= Q^{-1} \\ u_1 u_2 + u_2 u_3 + u_1 u_3 &= Q^{-1}\left(1 - \frac{r_1}{\epsilon}\right). \\ u_1 + u_2 + u_3 &= 1 - \frac{r_2}{\epsilon} \end{aligned} \quad (4.5)$$

As we already know that there is at least one positive equilibrium, it follows that there is a root of (4.4) for which the right sides of (4.3) are positive. Suppose that there are two or more positive equilibria. Then (4.4) must have more than one positive root such that the right sides of (4.3) are positive. In that case, (4.5) implies that all roots of (4.4) are positive and the sum of these roots equals $1 - \frac{r_2}{\epsilon}$. But the second of equations (4.3) implies that $u > 1 - \frac{r_2}{\epsilon}$ is necessary in order that $x_2 > 0$ and since we are assuming that there are at least two such roots of (4.4), together with a third positive root, we have a contradiction to the last of equations (4.5). We conclude that there is precisely one positive equilibrium of (4.2), which we denote by $\hat{x}$.

To see that $\hat{x}$ attracts all non-trivial positive orbits of (4.2) first note that if $x \gg 0$ then there exists $m, r > 0$ such that $rv \ll x \ll m1$ and where r is so small that $\phi_t(rv)$ converges to $\hat{x}$ (Theorem 3.3). Since $\phi_t(m1)$ also converges to $\hat{x}$ as $t \to \infty$ and $\phi_t(rv) \ll \phi_t(x) \ll \phi_t(m1)$ for $t > 0$, we conclude that $\phi_t(x) \to \hat{x}$ as $t \to \infty$. It is easy to see that any non-zero, nonnegative initial values for (4.2) give rise to a solution which is positive for all $t > 0$ and therefore, by the arguments above, this solution must converge to $\hat{x}$ as $t \to \infty$. $\square$

REMARK 4.1. If $\hat{J}$ denotes the Jacobian matrix of (4.2) at $x = \hat{x}$ then $s(\hat{J}) < 0$. In other words, $\hat{x}$ is asymptotically stable in the linear approximation. This fact will be useful later on and therefore we sketch the argument here. It is based on the inequality

$$\hat{J} = \begin{pmatrix} -\epsilon + r_1(1 - 2\hat{x}_1 K_1^{-1}) & \epsilon \\ \epsilon & -\epsilon + r_2(1 - 2\hat{x}_2 K_2^{-1}) \end{pmatrix}$$
$$< \begin{pmatrix} -\epsilon + r_1(1 - \hat{x}_1 K_1^{-1}) & \epsilon \\ \epsilon & -\epsilon + r_2(1 - \hat{x}_2 K_2^{-1}) \end{pmatrix}$$
$$= \bar{J}.$$

By Corollary 3.2 (3), $s(\hat{J}) < s(\bar{J})$. But $\bar{J}\hat{x} = 0$ since $\hat{x}$ is an equilibrium and therefore $s(\bar{J}) = 0$ by Corollary 3.2 (2). Hence, $s(\hat{J}) < 0$.

We write E_x for the corresponding equilibrium $(\hat{x}, 0)$ of (4.1). As an analogous result holds for the planar system obtained by setting $x = 0$ in (4.1), we let $E_y = (0, \tilde{y})$ be the unique equilibrium of (4.1) with $\tilde{y} \gg 0$ in $\mathbb{R}^2$. Let $E_0 = 0$ denote the trivial equilibrium of (4.1). We denote by f the vector field defined by (4.1) and let ϕ_t denote the corresponding flow.

Proposition 4.1 and monotonicity imply that (4.1) is dissipative. If $z = (x, y)$ is any point of $\mathbb{R}_+^4$ with $x, y > 0$, then $(0, y) \leq_m z \leq_m (x, 0)$ and therefore,

$$\phi_t((0, y)) \leq_m \phi_t(z) \leq_m \phi_t((x, 0)), \quad t > 0.$$

Since $\phi_t((0, y)) \to E_y$ and $\phi_t((x, 0)) \to E_x$ as $t \to \infty$, it follows that all positive orbits are attracted to the set

$$R \equiv [0, \hat{x}] \times [0, \tilde{y}] = \{z : E_y \leq_m z \leq_m E_x\}.$$

Also, observe that if $z = (x, y)$ satisfies $x, y > 0$ then $\phi_t(z) \gg 0$ for $t > 0$. This is intuitive since migration from an occupied patch to an unoccupied one is built into (4.1). In particular, z is also attracted to R.

Obviously, R contains E and, by Remark 1.1, if $z \in E$ with $z \gg 0$, then $E_y <_m z <_m E_x$ so strong monotonicity implies that $E_y \ll_m z \ll_m E_x$. Therefore, z belongs to the interior of R. Observe that R is positively invariant for (4.1) by monotonicity of the flow.

4. COMPETITION AND MIGRATION

Our next task is to examine the stability of the single-population equilibria E_x and E_y of (4.1). It suffices to consider only the stability of E_x since the stability of E_y is treated similarly. The Jacobian matrix, J_x, of f at E_x is given by:

$$J_x = \begin{pmatrix} \hat{J} & B \\ 0 & C \end{pmatrix}$$

where $\hat{J}$ is defined in Remark 4.1,

$$B = \begin{pmatrix} -r_1 a_1 \hat{x}_1 & 0 \\ 0 & -r_2 a_2 \hat{x}_2 \end{pmatrix}$$

and

$$C = \begin{pmatrix} -\delta + s_1(1 - b_1 \hat{x}_1) & \delta \\ \delta & -\delta + s_2(1 - b_2 \hat{x}_2) \end{pmatrix}.$$

By Remark 4.1, both eigenvalues of $\hat{J}$ are negative. Therefore, the stability of E_x is determined by the stability modulus, $s_x = s(C)$. The eigenvalues of C, one of which is s_x, are also real. An ugly formula can be given for s_x which we leave to the reader to construct.

An analogous result holds for the Jacobian matrix, J_y, of f at E_y. In this case, the stability of E_y is determined by the stability modulus, s_y, of the submatrix in the upper left of J_y. It is given explicitly by

$$s_y = s \begin{pmatrix} -\epsilon + r_1(1 - a_1 \tilde{y}_1) & \epsilon \\ \epsilon & -\epsilon + r_2(1 - a_2 \tilde{y}_2) \end{pmatrix}.$$

If $s_x < 0$ then E_x is asymptotically stable. If $s_x > 0$ then E_x is unstable and we have the following result.

THEOREM 4.2. *If $s_x > 0$, then one of the following holds.*
(i) *E_y attracts all solutions with initial data $z = (x, y)$ where $y > 0$. In this case, $s_y \leq 0$.*
(ii) *There exists a positive equilibrium E_* satisfying $E_y \ll_m E_* \ll_m E_x$. In this case, E_* attracts all solutions with initial data $z = (x, y)$ satisfying $E_* \leq_m z <_m E_x$ and $y > 0$.*

PROOF. $s(J_x) = s_x = s(C)$. We find a corresponding eigenvector (u, v) for J_x, $u, v \in \mathbb{R}^2$. By Corollary 3.2, there is an eigenvector $v \gg 0$ for C corresponding to s_x. Then, (u, v) is an eigenvector for J_x corresponding to s_x provided u satisfies

$$\hat{J}u + Bv = s_x u.$$

As noted above, both eigenvalues of $\hat{J}$ are negative and since $s_x > 0$, $s(\hat{J} - s_x) = s(\hat{J}) - s_x < 0$ so $\hat{J} - s_x I$ is invertible. Therefore,

$$u = -(\hat{J} - s_x I)^{-1} B v$$

is determined. By Corollary 3.2 (4), $-(\hat{J} - s_x I)^{-1} \gg 0$ and consequently, $u \ll 0$. Hence $(u, v) \ll_m 0$. By Theorem 3.3 and Remark 3.2, there exists $e > 0$ and $E_* \in E$ such that if $z_r = (\hat{x}, 0) + r(u, v)$ for $r \in (0, e]$ then

$$E_y \leq_m E_* \ll_m \phi_s(z_r) \ll_m \phi_t(z_r) \ll_m z_r \ll_m E_x$$

if $0 < t < s$ and $\phi_t(z_r) \to E_*$ as $t \to \infty$. There are two possibilities: (i) $E_* = E_y$, or (ii) $E_y \ll_m E_*$. We consider each case separately.

If $E_* = E_y$, then by Theorem 3.3, $\phi_t(z) \to E_y$ for all z satisfying $E_y \leq_m z \ll_m E_x$. In fact, if $z = (x, y) \in R$ and $x, y > 0$, then by Remark 1.2, $E_y \ll_m \phi_t(z) \ll_m E_x$ for all $t > 0$ and therefore $\phi_t(z) \to E_y$ as $t \to \infty$. In particular, $E = \{E_0, E_x, E_y\}$. If $z = (x, y) \in \mathbb{R}_+^4 \setminus R$, then, as noted above, $\phi_t(z) \to R$ as $t \to \infty$.

Now, (4.1) may be expressed as

$$x' = f_1(x) + O(|y|)$$
$$y' = Cy + O(|y|[|x - \hat{x}| + |y|]) \quad (4.6)$$

where f_1, defined in Proposition 4.1, is the right side of (4.2) and C is the two-by-two submatrix of J_x in the lower right. By Remark 3.1 and Remark 4.1, there is a vector $u \gg 0$ in $\mathbb{R}^2$ and $e > 0$ such that $f_1(\hat{x} + ru) \ll 0$ for all $r \in (0, e]$. From (4.6), if $p(r, s) = (\hat{x} + ru, sv)$ for small $r, s > 0$, where $Cv = s_x v$ as above, then at the point $p(r, s)$ we have

$$x' = f_1(\hat{x} + ru) + O(s)$$
$$y' = s[s_x v + O(r + s)].$$

Consequently, $x' \ll 0$ and $y' \gg 0$ for all small $r, s > 0$. By Proposition 3.2.1, $\phi_t(p(r, s)) \to E_y$ as $t \to \infty$. Since R attracts all positive orbits of (4.1), it follows that if $z = (x, y) \in \mathbb{R}_+^4$ and $y > 0$, then there exists $t_0 > 0$ and $r, s > 0$ such that $\phi_{t_0}(z) \ll_m p(r, s)$. As $\phi_{t+t_0}(z) \ll_m \phi_t(p(r, s))$ for $t > 0$, we conclude that $\phi_t(z) \to E_y$ as $t \to \infty$. This completes the proof if $E_* = E_y$. The proof of (2) is easier and we leave it to the reader. □

It is worth pointing out that Theorem 2.5.3, but not Theorem 3.3, can be applied to obtain a full monotone decreasing orbit of (4.1) connecting E_x to E_y in case (i) and E_x to E_* in case (ii) of Theorem 4.2. See Remark 3.4.

If both $s_x > 0$ and $s_y > 0$, then we might expect that coexistence of both populations occurs and that most orbits converge to a positive equilibrium. This is the content of the following result.

COROLLARY 4.3. *If $s_x > 0$ and $s_y > 0$, then there exist positive equilibria E_* and E_{**}, not necessarily distinct, satisfying*

$$E_y \ll_m E_{**} \leq_m E_* \ll_m E_x. \quad (4.7)$$

The order interval

$$O = \{z : E_{**} \leq_m z \leq_m E_*\}$$

attracts all solutions with initial condition $z = (x, y)$ satisfying $x, y > 0$. In particular, if $E_{**} = E_*$, then E_* attracts all solutions as above. An open and dense subset of initial conditions in $\mathbb{R}_+^4$ belong to positive orbits which converge to an equilibrium in O.

PROOF. The proof uses many of the ideas in the proof of Theorem 4.2 and so we use notation and arguments from that proof. E_* is obtained as a limit of a solution starting very close to E_x as in the previous proof. E_{**} is obtained from an entirely parallel argument as a limit of a solution starting very close to E_y. The inequality (4.7) follows immediately from monotonicity and the construction of these equilibria. Since all positive orbits are attracted to R and since E_* is also the limit of a solution starting at a point $p(r, s)$, as in Theorem 4.2, very near

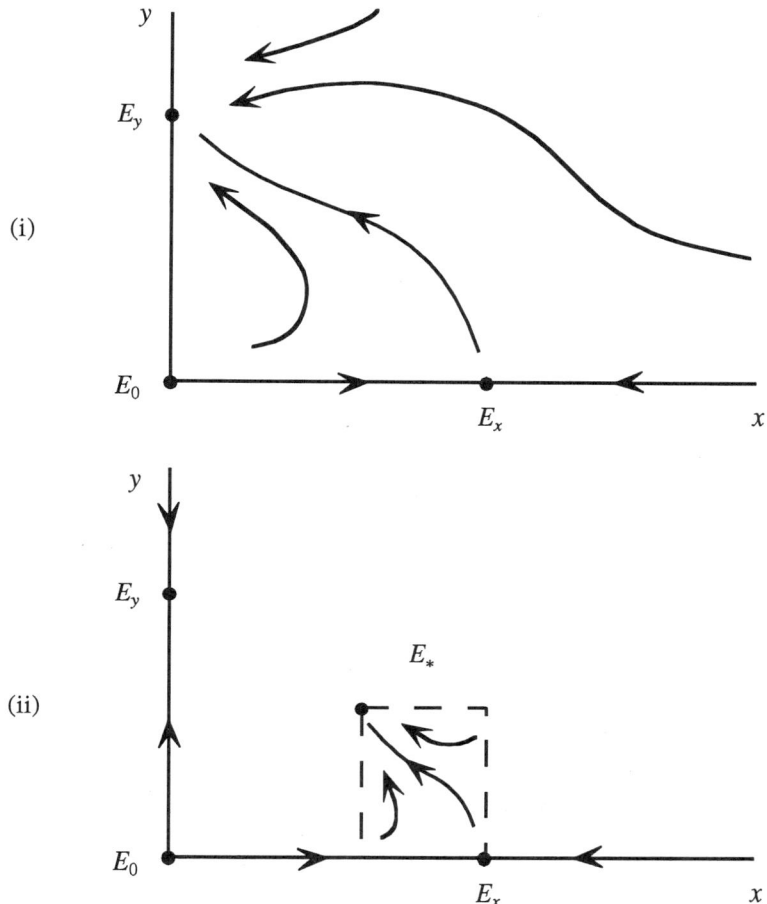

FIGURE 4.1. The two cases in Theorem 4.2.

E_x and E_{**} is the limit of a corresponding solution starting from a point $q(r,s)$ very near E_y (where the u component of $q(r,s)$ is positive and small and the v component is strictly larger than $\tilde{y}$), it follows that if $z = (x,y) \in \mathbb{R}_+^4$ and $x, y > 0$, then $q(r,s) \ll_m \phi_t(z) \ll_m p(r,s)$ for all large t and suitably small positive r, s. Therefore, O attracts all solutions starting at a point $z = (x,y)$ with $x, y > 0$. The final assertion follows from Theorem 1.2. □

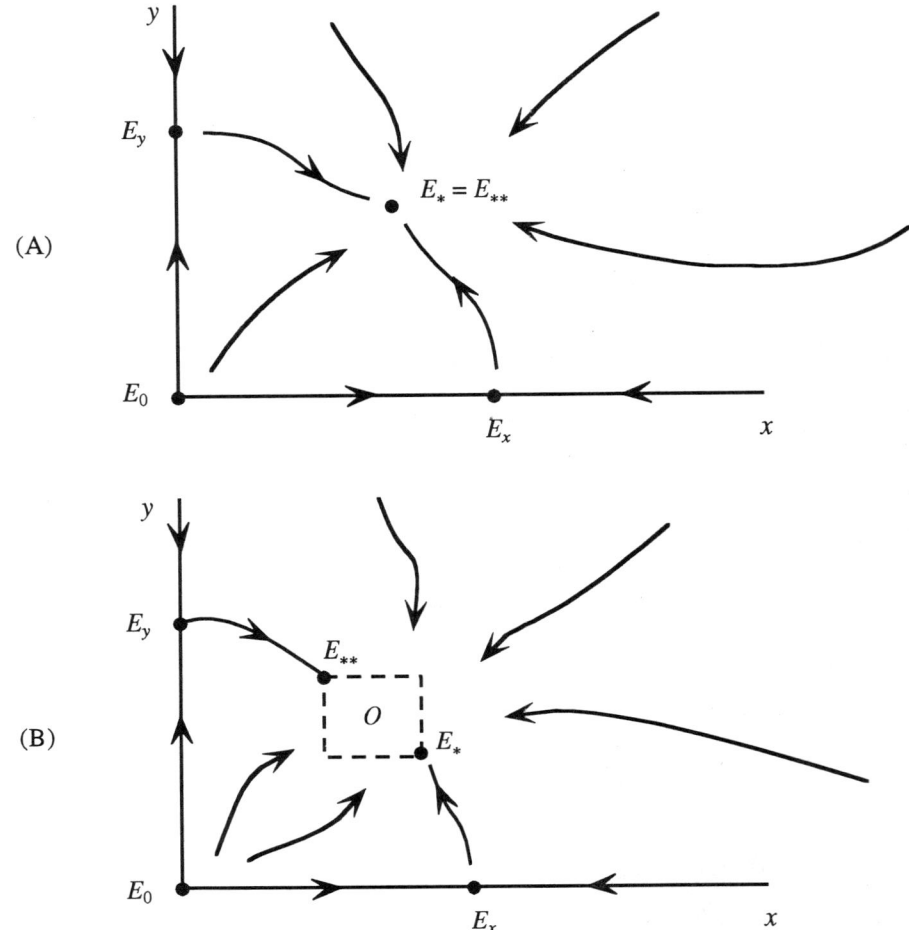

FIGURE 4.2. The phase portrait of (4.1) in case $E_* = E_{**}$ (A) and in case $E_{**} \ll_m E_*$ (B).

Figures 4.1 and 4.2 illustrate Theorem 4.2 and Corollary 4.3. Corollary 4.3 applies to the following example.

$$x_1' = 0.5(x_2 - x_1) + 2x_1(1 - 0.5x_1 - 0.5y_1)$$
$$x_2' = 0.5(x_1 - x_2) + 2x_2(1 - 0.5x_2 - 0.5y_2)$$
$$y_1' = 0.5(y_2 - y_1) + 5y_1(1 - 0.4y_1 - 0.2x_1)$$
$$y_2' = 0.5(y_1 - y_2) + 1.5y_2(1 - \frac{4}{3}y_2 - \frac{2}{3}x_2).$$

The results of this section hold without change for two populations occupying n discrete patches and that are able to migrate between patches. In this case, a crucial issue is the connectivity of the environment. The results described above

5. Smale's Construction

hold provided that members of each population can move from any given patch to any other given patch, possibly passing through other patches in route. This is just to say that the "diffusion" matrix for each population is irreducible. It need not be symmetric as in (4.1). Little is known about the case of more than two competitors. The resulting system is no longer monotone in this case.

5. Smale's Construction

In this section, we show that it is possible to imbed essentially arbitrary dynamics in a competitive or cooperative irreducible system. The aim is to construct a competitive system of the form

$$x'_i = x_i M_i(x), \ 1 \leq i \leq n, \tag{5.1}$$

in $\mathbb{R}^n_+$, where the M_i are smooth functions satisfying

$$\frac{\partial M_i}{\partial x_j} < 0 \tag{5.2}$$

for all i, j, and such that the standard $(n-1)$-simplex

$$\Sigma_n = \{x \in \mathbb{R}^n_+ : \sum x_i = 1\}$$

is an attractor on which arbitrary dynamics may be specified. All sums are understood to be from 1 to n. Equation (5.1) can be viewed as a simple model of competition between n populations.

In order to generate a dynamical system on Σ_n, let H denote the tangent space to Σ_n, that is,

$$H = \{x \in \mathbb{R}^n : \sum x_i = 0\},$$

and $h : \Sigma_n \to H$ be a smooth vector field on Σ_n. By smooth, we mean that partial derivatives of h of all orders exist and are continuous on Σ_n. We also assume that $h = (h_1, h_2, \ldots, h_n)$ has the form

$$h_i = x_i g_i(x)$$

where the g_i are smooth functions on Σ_n. Then

$$x'_i = h_i(x), \quad 1 \leq i \leq n, \tag{5.3}$$

generates a flow on Σ_n in the sense that Σ_n is invariant for the differential equations (5.3). The form of the h_i ensures that if $x_i(0) = 0$, then $x_i(t) \equiv 0$ so each lower dimensional simplex forming part of the boundary of Σ_n is invariant.

The goal is to construct a competitive system of the form (5.1) satisfying (5.2) such that its restriction to Σ_n is equivalent to (5.3). Let $S(x) = \sum x_j$ and let $p : \mathbb{R}_+ \to \mathbb{R}_+$ have continuous derivatives of all orders, be identically one in a neighborhood of $s = 1$, and vanish outside the interval $[1/2, 3/2]$. As g is a smooth vector field on Σ_n, it has a smooth extension to $\mathbb{R}^n_+$ which we denote by g in order to conserve notation (e.g., $P(\sum x_j)g(x/\sum x_j)/P(1)$ extends g to $\mathbb{R}^n_+$, where $P(u) = \int_0^u p(s)ds$). For $\eta > 0$, define

$$M_i(x) = 1 - S(x) + \eta p(\sum x_j) g_i(x), \quad 1 \leq i \leq n.$$

Then (5.2) holds for sufficiently small η since $p(\sum x_j)$ vanishes identically outside a compact subset of $\mathbb{R}^n_+$. Consider the system (5.1) with M as above. Clearly, $\mathbb{R}^n_+$

is positively invariant and the function S, evaluated along a solution $x(t)$ of (5.1), satisfies
$$\frac{d}{dt}S(x(t)) = S(x(t))(1 - S(x(t))),$$
since $\sum x_i g_i(x) = \sum h_i(x) = 0$. Consequently,
$$\sum x_i(t) \to 1, \quad t \to \infty,$$
unless, of course, $x(t) \equiv 0$. Therefore, Σ_n attracts all nontrivial solutions of (5.1). Furthermore, Σ_n is invariant for (5.1). In fact, on Σ_n, (5.1) becomes
$$x'_i = \eta h_i(x), \quad 1 \le i \le n.$$
Clearly then the dynamics of (5.1) restricted to Σ_n is equivalent, up to a change in time scale, to the dynamics generated by (5.3).

The construction above shows that any dynamics that can occur for a system (5.3) on the standard $(n-1)$-simplex in $\mathbb{R}_+^n$ can also occur in a competitive and irreducible system on $\mathbb{R}_+^n$ for which the simplex is an attractor. It seems clear that any compact attractor for a system of differential equations can be imbedded as an attractor in the standard simplex of appropriate dimension. This is roughly a converse result to Theorem 3.3.4. In particular, strange attractors and chaos can be expected in competitive systems of dimension exceeding three.

The construction also has implications for cooperative and irreducible systems since the time-reversed system corresponding to (5.1) is cooperative and irreducible in Int$\mathbb{R}_+^n$. Time-reversal, of course, makes the simplex a repeller for the cooperative system and therefore, any attractor with respect to the flow on the simplex is necessarily unstable for the flow of the cooperative system. Most orbits are attracted to the two equilibria, $x = 0$ and $x = \infty$. The simplex serves as the common boundary between the basins of attraction of these two equilibria. The construction tells us that Theorem 1.4.3 and Theorem 2.4.7 cannot be greatly improved. All orbits need not converge unless additional special hypotheses are assumed.

6. Remarks and Discussion

Theorem 1.1 on the strong monotonicity of the flow of a cooperative and irreducible system is due to Hirsch(1985) although (1.4) appears in Aronsson and Kellogg(1978). See also Martin(1978). Theorem 1.2 should be compared with Theorem 4.1 in Hirsch(1985) where it is shown that for Lebesgue almost all points having compact positive orbit closure, the solution converges to the set E. Our result requires more compactness than Hirsch's but then our conclusion is stronger in the sense that convergence, rather than quasiconvergence is generic.

It seems that there are no known examples of a cooperative system that is not irreducible and for which the conclusions of Theorem 1.2 fail to hold. Thus an important open problem is to either find such an example or prove that Theorem 1.2 is valid without the assumption of irreducibility. Jiang(1994) shows that if $n \le 4$ and every equilibrium is Liapunov stable, then every forward orbit of a cooperative system converges to an equilibrium.

For more on the mutualistic system (1.5), see Goh(1979) and Smith(1986c). While Theorem 1.2 cannot be immediately applied to (1.5), it can be asserted that of those points $x \gg 0$ for which the positive orbit is bounded and $\omega(x) \gg 0$, almost all, with respect to Lebesgue measure, converge to an equilibrium. This follows by applying Theorem 4.1 of Hirsch(1985) to the system obtained from (1.5) after the transformation $y_i = \log x_i$. If $r \gg 0$ and $s(A) < 0$, then there is a unique equilibrium $e \gg 0$ which attracts all positive orbits (Smith(1986c)).

Our presentation of the example of section two was motivated by work of Selgrade(1980,1982). The latter paper of Selgrade shows that an unstable Hopf bifurcation can occur from the middle of three equilibria in (2.1). Consequently, Theorem 2.3.2 is sharp. See also Smith(1987b), where further applications of monotone systems theory to this class of models are made, and the references therein. There is a huge literature on these protein synthesis models.

The ideas used in Theorem 3.3 have appeared many times in the literature. A partial list of references include Hess(1991), Smith(1986a,c,d), Smith et al(1991), Takeuchi and Lu(1993). Our proof of the last part of Theorem 3.3 borrows techniques from Selgrade(1983) but is different. Results similar to Theorem 3.3 appear in Matano(1984), Selgrade(1983), Smith(1986b), Poláčik(1990), Dancer and Hess(1991). The result of Selgrade(1983) is specifically formulated for cooperative systems of ordinary differential equations and establishes the uniqueness of the heteroclinic orbit. Selgrade's result is further improved in Jiang(1992). The results of the other authors apply to infinite dimensional systems.

Proposition 3.4 appears in Smith(1986e) where it is also shown that there is an invariant cylinder-like "most unstable" manifold associated with a periodic orbit of an irreducible cooperative system. This result was further exploited in Smith and Waltman(1987) for three dimensional Kolmogorov-type competitive systems.

The competition and migration model of section 4 is patterned after results of Takeuchi(1989) and Takeuchi and Lu(1994), although our notation is different. The mathematics is isomorphic to that used in the analysis of a model of microbial competition in a continuous culture device called the gradostat. See Jager et al(1987) and Smith et al(1991). A model of two competing subcommunities consisting of mutualists, Smith(1986d), contains a similar analysis.

The generalization of the results of section four to include n patches is straightforward except for the uniqueness of each single-population equilibrium. The uniqueness follows immediately from the main result in Smith(1986a) provided each diffusion matrix is irreducible. The proof of uniqueness can also be given by modifying slightly the uniqueness argument of Theorem 8.2.1. See also Takáč(1990) for an extension of the results in Smith(1986a) to infinite dimensional systems.

The examples (2.1) and (4.1) illustrate a common feature of dissipative monotone systems. Such systems have a compact attractor A which attracts all orbits. If there exists $x \in X$ such that $x \ll A$ then since $\omega(x) \subset A$, there exists a $t_0 > 0$ such that $\phi_{t_0}(x) \gg x$. By the Convergence Criterion there exists $u \in E$ such that $\phi_t(x) \to u$ as $t \to \infty$. Since A is invariant and attracts all orbits, $u \leq A$ and $u \in A$. Furthermore, u is asymptotically stable from below since it attracts the set $\{z \in X : x \leq z \leq u\}$. If also, there exists $y \in X$ such that $A \ll y$, then by similar reasoning, there exists $v \in E \cap A$ such that $A \leq v$ and v is asymptotically stable from above. Thus, $A \subset [u,v]$. Hirsch(1984) shows that an attractor for a monotone flow always contains an equilibrium and if it contains only one, then every orbit attracted to the attractor converges to the equilibrium. Furthermore, if the semiflow is strongly monotone, then the attractor contains an *order stable* equilibrium. An order stable equilibrium is an equilibrium which is stable with respect to the topology generated by the open order intervals $\{x \in X : a \ll x \ll b\}$. See Jiang(1991) for additional results in this direction.

The construction of section 5 is, of course, due to Smale(1976). This paper has had a profound influence on the theory of monotone systems although it was originally intended to chasten mathematical biologists who were confident that competitive systems could only have simple dynamics. It still provides a ready source of counterexamples to naive conjectures regarding monotone systems.

An example of a strange attractor in a four dimensional Lotka-Volterra competition model appears in Arneodo et al(1982).

CHAPTER 5

Cooperative Systems of Delay Differential Equations

0. Introduction

The aim of the present chapter is to apply the theory developed in Chapters 1 and 2 to differential equations containing delayed arguments. Such equations are often referred to as delay differential equations or functional differential equations. Since delay differential equations contain ordinary differential equations as a special case (when all delays are zero), the treatment here follows that of the previous two chapters and the results of this chapter generalize many of the results of Chapters 3 and 4. The main difference is that a delay differential equation generally can't be solved backward in time and therefore there is not a well-developed theory of competitive systems with delays.

Delay differential equations generate infinite dimensional dynamical systems and there are several choices of state space. We restrict attention here to equations with bounded delays and follow the most well-developed theory (see Hale(1977)). The Remarks and Discussion section contains references to various extensions. If r denotes the maximum delay appearing in the equation, then the space $C = C([-r, 0], \mathbb{R}^n)$ is a natural choice of state space. It contains the cone of functions which map $[-r, 0]$ into $\mathbb{R}^n_+$. The chapter begins by identifying sufficient conditions on the delay differential equation for the semiflow to be monotone with respect to this ordering. This condition, called the quasimonotone condition, reduces to the Kamke condition when no delays are present. However, a slight change in the state space is required in order to show that the semiflow is eventually strongly monotone. In section 3, the class of cooperative and irreducible delay differential equations are introduced. These systems generate an eventually strongly monotone semiflow for which the generic orbit converges to equilibrium.

In section 2, we develop conditions guaranteeing the existence of positively invariant sets which are rectangles or hypercubes in $\mathbb{R}^n$. When these conditions are satisfied, certain quasimonotone comparison systems can be associated with the delay equation. The analog of Proposition 3.2.1 on the existence of monotone solutions is proved. Finally, the method of contracting rectangles for obtaining global convergence of an equilibrium is described.

In Chapter 4 we saw that the Perron-Frobenius Theorem played a crucial role in stability considerations for cooperative and irreducible systems of ordinary differential equations. It implies that the dominant eigenvalue of the Jacobian matrix at an equilibrium is real and simple and there is a corresponding positive eigenvector. It should come as no surprise that the Krein-Rutman Theorem, the infinite dimensional analog of the Perron-Frobenius Theorem, plays a similar role in stability considerations for cooperative systems of delay equations. The remarkable conclusion that can be drawn from our analysis in section 5 is that an equilibrium of a cooperative system of delay differential equations has the same stability properties as the cooperative system of ordinary differential equation obtained by setting all delays to zero.

Applications are made to the biochemical control loop introduced in Chapter 4 where delays in the catalytic terms are introduced and to a scalar delay equation that arises in applications to many fields. Section 7 is devoted to the application of the contracting rectangles idea to the Lotka-Volterra system of n competing populations where time delays are incorporated into the intraspecific and interspecific competition terms. This system does not generate a monotone semiflow unless $n = 2$, but we give conditions for the existence of a globally stable coexistence equilibrium for arbitrary n.

For the reader unfamiliar with delay equations, we offer a brief review in the remainder of this introduction. Much can be learned by considering the simple equation

$$x'(t) = -x(t) + h(x(t-r)), \quad t \geq 0,$$

where h is a continuous function. A bit of reflection will convince the reader that $x(t)$ must be prescribed on the interval $[-r, 0]$ in order that it is determined for $t \geq 0$. A natural space of initial conditions is the space of continuous functions on $[-r, 0]$, which we denote by $C \equiv C([-r, 0], \mathbb{R}^n)$, where $n = 1$ in this case. C is a Banach space with the usual uniform norm $|\phi| = \sup\{|\phi(\theta)| : -r \leq \theta \leq 0\}$. If $\phi \in C$ is given, then it is easy to see that the equation has a unique solution $x(t)$ for $t \geq 0$ satisfying

$$x(\theta) = \phi(\theta), \quad -r \leq \theta \leq 0.$$

In fact, for $0 \leq t \leq r$, $x(t)$ must satisfy

$$x'(t) = -x(t) + h(\phi(t-r)), \quad x(0) = \phi(0)$$

so $x(t)$ can be determined on this interval by an integration. The method of steps consists of successively applying this argument on the intervals $[(n-1)r, nr]$ for $n = 1, 2, \ldots$. If the space of initial conditions is taken to be C, then it is reasonable to view this space as the "phase space" for the equation, rather than the space $\mathbb{R}$ to which $x(t)$ belongs. Therefore we are led to construct, from the solution $x(t)$, an element of the space C to call the state of the system at time t. It should have the property that it uniquely determines $x(s)$ for $s \geq t$. The natural choice is $x_t \in C$, defined by

$$x_t(\theta) = x(t+\theta), \quad -r \leq \theta \leq 0.$$

Then, $x_0 = \phi$ and $x_t(0) = x(t)$. **Figure 0.1** illustrates this choice.

More complicated examples, such as

$$x'(t) = \int_{-r}^{0} x(t+\theta) d\eta(\theta)$$

where $\eta \in BV = BV[-r, 0]$, the space of functions on $[-r, 0]$ of bounded variation, require a more general approach.

A general autonomous functional differential equation is denoted by

$$x'(t) = f(x_t) \tag{0.1}$$

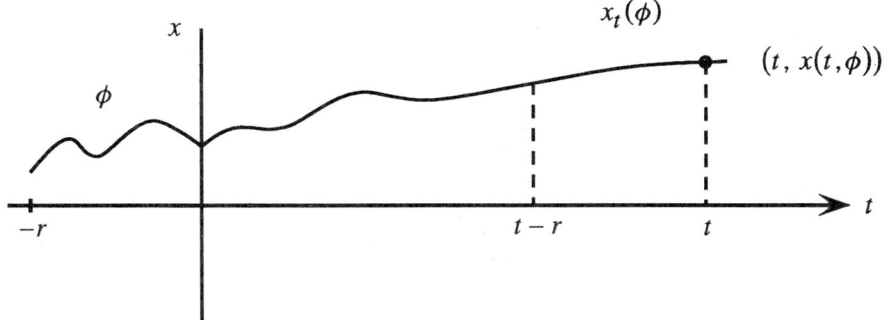

FIGURE 0.1. $x_t(\phi)$ is obtained by translating the graph of $x(t,\phi)$ on the interval $[t-r,t]$ to the interval $[-r,0]$.

where $f : D \to \mathbb{R}^n$, D is an open subset of C and f is continuous. In the two examples above, f is given by

$$f(\phi) = -\phi(0) + h(\phi(-r)),$$

and

$$f(\phi) = \int_{-r}^{0} \phi(\theta) d\eta(\theta)$$

for $\phi \in C$. Observe that (0.1) includes the system of ordinary differential equations

$$x' = g(x)$$

where $g : \mathbb{R}^n \to \mathbb{R}^n$, as a special case. Simply let $f(\phi) = g(\phi(0))$ so that $f(x_t) = g(x_t(0)) = g(x(t))$.

It will always be assumed that (0.1), together with the initial condition $x_0 = \phi \in D$ has a unique, maximally defined solution, denoted by $x(t,\phi)$, on an interval $[0,\sigma)$. The state of the system is denoted by $x_t(\phi)$ to emphasize the dependence on the initial data. Uniqueness of solutions holds if, for example, f is Lipschitz on compact subsets of D. If uniqueness of solutions of initial value problems holds, then the map $(t,\phi) \to x_t(\phi)$ is continuous. Therefore, a (local) semiflow on D can be defined by

$$\Phi_t(\phi) = x_t(\phi).$$

In contrast to the case of ordinary differential equations, $x(t,\phi)$ cannot usually be defined for $t \leq 0$ as a solution of (0.1) and consequently, Φ_t need not be one-to-one. Hale(1977) is a standard reference for functional differential equations.

The equilibria of (0.1) are those $\phi \in D$ such that $x_t(\phi) = \phi$ for all $t \geq 0$. Evaluating this last identity at $\theta = 0$ gives $x(t) = \phi(0)$ for all $t \geq 0$ so ϕ must be a

constant function and $f(\phi) = 0$. Thus the set of equilibria is given by
$$E = \{\hat{x} \in D : x \in \mathbb{R}^n \text{ and } f(\hat{x}) = 0\},$$
where for $x \in \mathbb{R}^n$ we write $\hat{x}$ for the element of C satisfying $\hat{x}(\theta) \equiv x$. In particular, the equilibria of our first example consist of $\hat{x}$ for those x which are solutions of $h(x) = x$.

1. The Quasimonotone Condition

The Banach space C contains the cone
$$C_+ = \{\phi \in C : \phi(\theta) \geq 0, \ -r \leq \theta \leq 0\}.$$
The usual notation $\leq, <, \ll$ will be used for the various order relations on C generated by C_+. In particular, $\phi \leq \psi$ holds in C if and only if $\phi(s) \leq \psi(s)$ holds in $\mathbb{R}^n$ for every $s \in [-r, 0]$. The same notation will also be used for the various order relations on $\mathbb{R}^n$ but hopefully the context will alert the reader to the appropriate meaning. It will be convenient to have notation for the natural imbedding of $\mathbb{R}^n$ into C. If $x \in \mathbb{R}^n$, let $\hat{x} \in C$ be the constant function equal to x for all values of its argument. Finally, let e_i denote the standard basis vectors for $\mathbb{R}^n$, $1 \leq i \leq n$.

Our immediate aim is to identify sufficient conditions on f for the semiflow Φ to be a monotone semiflow. The following condition should seem natural since it generalizes the Kamke condition for ordinary differential equations. We refer to it as the *quasimonotone condition*.

(Q) Whenever $\phi \leq \psi$ and $\phi_i(0) = \psi_i(0)$ holds for some i, then $f_i(\phi) \leq f_i(\psi)$.

(Q) is sufficient to guarantee that Φ is a monotone semiflow. Actually, for the same reasons as in Chapter 3, it is convenient to consider the nonautonomous equation
$$x'(t) = f(t, x_t) \tag{1.1}$$
where $f : \Omega \to \mathbb{R}^n$ is continuous on Ω, an open subset of $\mathbb{R} \times C$. Given $(t_0, \phi) \in \Omega$, we write $x(t, t_0, \phi, f)$ and $x_t(t_0, \phi, f)$ for the maximally defined solution and state of the system at time t which satisfies (1.1) for $t \geq t_0$ and $x(t_0 + \theta) = \phi(\theta)$ for $-r \leq \theta \leq 0$ (equivalently, $x_{t_0} = \phi$). We assume this solution is unique, which will be the case if f is Lipschitz in its second argument on each compact subset of Ω. We drop the last argument f from $x(t, t_0, \phi, f)$ when no confusion over which f is being considered will result. f is said to satisfy (Q) if for each t such that $\Omega_t \equiv \{\phi \in C : (t, \phi) \in \Omega\}$ is nonempty, $f(t, \bullet)$ satisfies (Q) on Ω_t.

The next Theorem not only establishes the desired monotonicity of the semiflow Φ but also allows comparisons of solutions between related functional differential equations. It thus represents a generalization of Proposition 3.1.1 and contains Remark 1.2 of Chapter 3 as a special case.

THEOREM 1.1. *Let $f, g : \Omega \to \mathbb{R}^n$ be continuous, Lipschitz on each compact subset of Ω, and assume that either f or g satisfies (Q). Assume also that $f(t, \phi) \leq g(t, \phi)$ for all $(t, \phi) \in \Omega$. If $(t_0, \phi), (t_0, \psi) \in \Omega$ satisfy $\phi \leq \psi$, then*
$$x(t, t_0, \phi, f) \leq x(t, t_0, \psi, g)$$
holds for all $t \geq t_0$ for which both are defined.

1. THE QUASIMONOTONE CONDITION

PROOF. Assume that f satisfies (Q), a similar argument holds if g satisfies (Q). Let $e = (1, 1, \ldots, 1)$, $g_\epsilon(t, \phi) = g(t, \phi) + \epsilon e$ and $\psi_\epsilon = \psi + \epsilon \hat{e}$, for $\epsilon \geq 0$. If $x(t, t_0, \psi, g)$ is defined on $[t_0 - r, t_1]$ for some $t_1 > t_0$, then $x(t, t_0, \psi_\epsilon, g_\epsilon)$ is also defined on this same interval for all sufficiently small positive ϵ and

$$x(t, t_0, \psi_\epsilon, g_\epsilon) \to x(t, t_0, \psi, g), \quad \epsilon \to 0,$$

for $t \in [t_0, t_1]$ by Hale(1977), Theorem 2.2. We will show that $x(t, t_0, \phi, f) \ll x(t, t_0, \psi_\epsilon, g_\epsilon)$ on $[t_0 - r, t_1]$ for small positive ϵ. The result will then follow by letting $\epsilon \to 0$. If the assertion above were false for some ϵ, then there exists $s \in (t_0, t_1]$ such that $x(t, t_0, \phi, f) \ll x(t, t_0, \psi_\epsilon, g_\epsilon)$ for $t_0 \leq t < s$ and $x_i(s, t_0, \phi, f) = x_i(s, t_0, \psi_\epsilon, g_\epsilon)$ for some i. Clearly, $x'_i(s, t_0, \phi, f) \geq x'_i(s, t_0, \psi_\epsilon, g_\epsilon)$ must hold. But

$$\begin{aligned} x'_i(s, t_0, \psi_\epsilon, g_\epsilon) &= g_i(s, x_s(t_0, \psi_\epsilon, g_\epsilon)) + \epsilon \\ &> f_i(s, x_s(t_0, \psi_\epsilon, g_\epsilon)) \\ &\geq f_i(s, x_s(t_0, \phi, f)) \\ &= x'_i(s, t_0, \phi, f), \end{aligned}$$

where the last inequality follows from (Q). This contradiction implies that no such s can exist and proves the assertion. $\square$

In the case of the autonomous system (0.1), taking $f = g$ in Theorem 1.1 implies that $x_t(\phi) \leq x_t(\psi)$ for $t \geq 0$ such that both solutions are defined. In contrast to Proposition 3.1.1, if $\phi < \psi$ we cannot generally conclude that $x(t, \phi) < x(t, \psi)$ or $x_t(\phi) < x_t(\psi)$ since Φ_t is not generally one-to-one.

As an example, consider the delay differential equation with only a single delay given by

$$x'(t) = g(x(t), x(t - r)) \tag{1.2}$$

where $r > 0$ and $g : \mathbb{R}^n \times \mathbb{R}^n \to \mathbb{R}^n$ is continuously differentiable. Then (Q) holds provided that $g(x, y)$ satisfies

$$\frac{\partial g_i}{\partial x_j} \geq 0, \quad i \neq j,$$

and

$$\frac{\partial g_i}{\partial y_l} \geq 0, \quad \text{for all } i, l.$$

In particular, if (1.2) is a scalar equation ($n = 1$), then $g(x, y)$ must be nondecreasing in y for each fixed x.

As a second example, consider the general nonautonomous linear system

$$x'(t) = L(t)x_t \tag{1.3}$$

where $L : \mathbb{R} \to L(C, \mathbb{R}^n)$ is continuous and $L(C, \mathbb{R}^n)$ is the space of bounded linear maps from C into $\mathbb{R}^n$. $L(t)$ satisfies (Q) if and only if the following holds where $L_i(t)\phi$ denotes the i-th component of $L(t)\phi$.

(K) Whenever $\phi \geq 0$ and $\phi_i(0) = 0$ then $L_i(t)\phi \geq 0$.

It will prove to be convenient to have a representation of $L(t)$ in terms of Borel measures.

LEMMA 1.2. (K) *holds if and only if there exists* $a_i(t) \in \mathbb{R}$ *for* $1 \leq i \leq n$ *and positive Borel measures* $\eta_{ij}(t)$ *for* $1 \leq i, j \leq n$ *such that*

$$L_i(t)\phi = a_i(t)\phi_i(0) + \sum_{j=1}^{n} \int_{-r}^{0} \phi_j(\theta) d_\theta \eta_{ij}(t, \theta), \tag{1.4}$$

and $\eta_{ii}(t)\{0\} = 0$. *Moreover, if* (K) *holds then the representation* (1.4) *is unique and* $a_i(t)$ *and* $\eta_{ij}(t)$ *are continuous functions of* t.

PROOF. If (K) holds, then for each t and $j \neq i$, the map $\alpha \to L_i(t)(\alpha e_j)$, where $\alpha \in C([-r, 0], \mathbb{R})$ is a positive linear functional. By the Riesz Representation Theorem, Rudin(1966), this functional is represented by a unique positive measure $\eta_{ij}(t)$. Taking $i = j$ in the previous argument gives rise to a linear functional x^* on $C([-r, 0], \mathbb{R})$ which, by (K), has the property that $x^*(\alpha) \geq 0$ whenever $\alpha \geq 0$ and $\alpha(0) = 0$. x^* can be represented by

$$x^*(\alpha) = \int_{-r}^{0} \alpha(\theta) d\mu(\theta)$$

where $\mu \in BV[-r, 0]$ is continuous on the left at each point of $(-r, 0)$. Let α_n be a sequence of nonnegative continuous functions on $[-r, 0]$ which vanish at 0, satisfy $\alpha_n(\theta) \leq 1$ on $[-r, 0]$ and which converge pointwise to χ, the characteristic function of the set $[a, b]$ where $-r \leq a < b < 0$. By a limiting argument, it follows that

$$\mu(b) - \mu(a) = \int_{-r}^{0} \chi(\theta) d\mu(\theta) \geq 0.$$

Therefore, $\mu(\theta)$ is nondecreasing on $[-r, 0)$. Let $a = \mu(0) - \mu(0-)$ and define η on $[-r, 0]$ by $\eta(\theta) = \mu(\theta)$ for $\theta < 0$ and $\eta(0) = \mu(0-)$. Then $\eta(\theta)$ is nondecreasing on $[-r, 0]$ and

$$x^*(\alpha) = a\phi(0) + \int_{-r}^{0} \alpha(\theta) d\eta(\theta)$$

for $\alpha \in C([-r, 0], \mathbb{R})$. If we let η_{ii} denote the measure defined by η, then η_{ii} is a positive measure satisfying $\eta_{ii}(\{0\}) = 0$. It is clear that this representation of x^* is unique since μ is unique modulo addition of a scalar which does not affect a.

Conversely, if the representation (1.4) holds then it is easy to see that (K) follows immediately. The continuity of a_i and η_{ij} follows from the continuity of $L(t)$. □

As a consequence of Lemma 1.2, we can express $L(t)$ as

$$L(t)\phi = D(t)\phi(0) + \bar{L}(t)\phi \tag{1.5}$$

where $D(t) = \mathrm{diag}(a_1(t), a_2(t), \ldots, a_n(t))$ and

$$\bar{L}_i(t)\phi = \sum_{j=1}^{n} \int_{-r}^{0} \phi_j(\theta) d_\theta \eta_{ij}(t, \theta).$$

$\bar{L}(t)$ has the property that $\bar{L}(t)\phi \geq 0$ if $\phi \geq 0$.

The next result implies that whenever a component of a solution of (1.3) gets turned on, then it stays on.

LEMMA 1.3. *Let (K) hold for (1.3), $\phi \in C_+$ and $x(t) = x(t, t_0, \phi)$, $t \geq t_0$ satisfy (1.3). If $x_i(t_1) > 0$ for some $t_1 \geq t_0$, then $x_i(t) > 0$ for $t \geq t_1$.*

PROOF. Since $x_t \geq 0$ and (K) holds, it follows that

$$\begin{aligned} x_i'(t) &= L_i(t) x_t \\ &\geq a_i(t) x_i(t) + \bar{L}_i(t) x_t \\ &\geq a_i(t) x_i(t). \end{aligned}$$

A standard comparison argument implies that

$$x(t) \geq x(t_1) \exp\left(\int_{t_1}^{t} a_i(s) ds \right) > 0$$

for $t \geq t_1$. □

2. Positively Invariant Sets, Monotone Solutions, and Contracting Rectangles

It is very important to have sufficient conditions for certain subsets of C to be positively invariant for a functional differential equation that does not necessarily satisfy the quasimonotone condition (Q). To provide such conditions is only one of the aims of the present section. Here, we will focus exclusively on sufficient conditions for certain order intervals, $[\hat{a}, \hat{b}]$ to be positively invariant where $a, b \in \mathbb{R}^n$ or are infinite. These sufficient conditions lead naturally to sufficient conditions for solutions of quasimonotone functional differential equations to be monotone increasing or decreasing, just as in Proposition 3.2.1 for cooperative ordinary differential equations. However, the most important results of this section apply to delay differential equations which do not necessarily satisfy the quasimonotone condition. Although the condition (Q) is quite special, we show that to any delay differential equation with a positively invariant order interval $[a, b]$, there exist two auxiliary delay differential equations $x'(t) = h(x_t)$ and $x'(t) = H(x_t)$, where h, H satisfy (Q), such that the solutions of (0.1) can be directly compared to those of the auxilliary systems. Such comparisons sometimes lead to the result that all solutions of (0.1) corresponding to initial conditions in the order interval converge to a unique equilibrium. Therefore, quasimonotone systems arise naturally in various comparison arguments involving systems that are not quasimonotone.

We begin by asking when the cone C_+ is positively invariant for (0.1). Since no additional work is involved in obtaining our first result for nonautonomous systems, we consider the system

$$x'(t) = f(t, x_t) \tag{2.1}$$

where $f : \mathbb{R} \times D \to \mathbb{R}^n$ is continuous and $D \subset C$ is open. It is also assumed that f is Lipschitz in its second argument on each compact subset of $\mathbb{R} \times D$ so that initial value problems associated with (2.1) have unique solutions. If $C_+ \cap D$ is nonempty, we can ask for conditions on f which ensure that $C_+ \cap D$ is positively invariant. Such conditions are given in the following result.

THEOREM 2.1. *Assume that whenever $\phi \in D$ satisfies $\phi \geq 0$, $\phi_i(0) = 0$ for some i and $t \in \mathbb{R}$, then $f_i(t, \phi) \geq 0$. If $\phi \in D$ satisfies $\phi \geq 0$ and $t_0 \in \mathbb{R}$, then $x(t, t_0, \phi) \geq 0$ for all $t \geq t_0$ in its maximal interval of existence.*

PROOF. The proof is very similar to that of Theorem 1.1 and we will use the notation developed there. If $f_\epsilon(t,\phi) = f(t,\phi) + \epsilon\hat{e}$, then we will show that $x(t) = x(t,t_0,\phi,f_\epsilon) \geq 0$ whenever $\phi \geq 0$ and $\epsilon > 0$. The result will then follow by letting $\epsilon \to 0$ as in the proof of Theorem 1.1. Now, $x(t_0) = \phi(0) \geq 0$ and if $x_i(t_0) = 0$ then, by hypothesis, $x'_i(t_0) = f_i(t_0,\phi) + \epsilon > 0$. Therefore, if the result were false, we can assume that there exists $t_1 > 0$ such that $x(t) \gg 0$ for $t_0 < t < t_1$ and $x_i(t_1) = 0$ for some i. It would follow that $x'_i(t_1) \leq 0$. However, $x_{t_1} \in C_+$ and so, by hypothesis,
$$x'_i(t_1) = f_i(t_1, x_{t_1}) + \epsilon > 0.$$
This contradiction proves the result. □

Some more or less obvious extensions of Theorem 2.1 are pointed out below.

REMARK 2.1. Let $h \in \mathbb{R}^n$ and $[\hat{h}, \infty] = \{\phi \in C : \phi \geq \hat{h}\}$. $[\hat{h}, \infty] \cap D$ is positively invariant for (2.1) provided that whenever $\phi \in D$, $\phi \geq \hat{h}$, $t \in \mathbb{R}$ and $\phi_i(0) = h_i$ for some i, then $f_i(t,\phi) \geq 0$. The simple proof follows from Theorem 2.1 by a change of variables in (2.1). Let $[-\infty, \hat{h}] = \{\phi \in C : \phi \leq \hat{h}\}$. It is positively invariant provided that whenever $\phi \in D$, $\phi \leq \hat{h}$, $t \in \mathbb{R}$ and $\phi_i(0) = h_i$ for some i, then $f_i(t,\phi) \leq 0$. Finally, the order interval $[\hat{h},\hat{k}] \cap D$, for $h,k \in \mathbb{R}^n$ satisfying $h \ll k$, is positively invariant provided that whenever $\phi \in [\hat{h},\hat{k}] \cap D$, $t \in \mathbb{R}$ and $\phi_i(0) = h_i$ ($\phi_i(0) = k_i$) for some i, then $f_i(t,\phi) \geq 0$ ($f_i(t,\phi) \leq 0$).

The next result provides sufficient conditions for the existence of monotone solutions of a quasimonotone autonomous system that converge to an equilibrium point. It contains part of Proposition 3.2.1 for cooperative systems of ordinary differential equations as a special case.

COROLLARY 2.2. Let f in (0.1) satisfy (Q). If $h \in \mathbb{R}^n$ is such that $\hat{h} \in D$ and $f(\hat{h}) \geq 0$ ($f(\hat{h}) \leq 0$), then $x(t,\hat{h})$ is nondecreasing (nonincreasing) in $t \geq 0$. If the positive orbit of $\hat{h}$ has compact closure in D, then there exists $k \geq h$ ($k \leq h$) such that $x(t,\hat{h}) \to k$ as $t \to \infty$.

PROOF. Suppose that $f(\hat{h}) \geq 0$, the proof in the other case is similar. If $\phi \geq \hat{h}$ and $\phi_i(0) = h_i$, then by (Q), we conclude that $f_i(\phi) \geq f_i(\hat{h}) \geq 0$. Therefore, by Remark 2.1, $[\hat{h}, \infty] \cap D$ is positively invariant for (0.1). In particular, $x_t(\hat{h}) \geq \hat{h}$ for $t \geq 0$. But then Theorem 1.1 implies that $x_{t+s}(\hat{h}) \geq x_s(\hat{h})$ for $t, s \geq 0$. Equivalently, $\hat{h} \leq x_s(\hat{h}) \leq x_t(\hat{h})$ whenever $0 \leq s \leq t$. Evaluating each function at $\theta = 0$ in the previous inequality yields the conclusion that $x(t,\hat{h})$ is nondecreasing in t. If the positive orbit of $\hat{h}$ has compact closure in D then $x_t(\hat{h})$ must converge to an equilibrium. Therefore, there is a $k \geq h$ such that $x_t(\hat{h}) \to \hat{k}$ as $t \to \infty$. □

Observe that if $f(\hat{h}) \geq 0$ ($f(\hat{h}) \leq 0$) then $\hat{h}$ is a sub (super)-equilibrium for the semiflow Φ generated by (0.1). See section 5 of Chapter 2.

The quasimonotone condition (Q) is quite special but now we describe how it can arise naturally from the consideration of a system (0.1) which does not necessarily satisfy (Q) but which has a positively invariant order interval
$$\Sigma = [\hat{a},\hat{b}],$$

where $a, b \in \mathbb{R}^n$. That is, we assume that

$$\phi \in \Sigma \text{ and } \phi_i(0) = a_i \ (\phi_i(0) = b_i) \Rightarrow f_i(\phi) \geq 0 \ (f_i(\phi) \leq 0), \tag{2.2}$$

and we assume that f satisfies a Lipschitz condition on Σ. More precisely, there exists $L > 0$ such that

$$|f(\psi) - f(\phi)| \leq L|\psi - \phi|, \quad \psi, \phi \in \Sigma. \tag{2.3}$$

Throughout this section, $|a| = \max_i |a_i|$ for $a \in \mathbb{R}^n$. Define comparison functions $h, H : \Sigma \to \mathbb{R}^n$ for f as follows. Given $\phi \in \Sigma$ and $1 \leq i \leq n$, let

$$\begin{aligned} h_i(\phi) &= \inf\{f_i(\psi) : \phi \leq \psi \leq \hat{b} \text{ and } \psi_i(0) = \phi_i(0)\} \\ H_i(\phi) &= \sup\{f_i(\psi) : \hat{a} \leq \psi \leq \phi \text{ and } \psi_i(0) = \phi_i(0)\} \end{aligned} \tag{2.4}$$

Since f satisfies (2.3), it is bounded on Σ so the infimum and supremum defining h and H are finite. The relationship of h and H to f is revealed in the next result. H is the smallest quasimonotone function larger than or equal to f and h is the largest quasimonotone function less than or equal to f.

PROPOSITION 2.3. *H and h satisfy (2.2), (2.3) and (Q) on Σ and*

$$h(\phi) \leq f(\phi) \leq H(\phi), \quad \phi \in \Sigma. \tag{2.5}$$

If f satisfies (Q), *then $f = h = H$ on Σ.*

PROOF. The inequality (2.5) is immediate from the definitions of h and H. If $\phi \in \Sigma$ satisfies $\phi_i(0) = a_i$ for some i, then $H_i(\phi) \geq f_i(\phi) \geq 0$, where the last inequality holds by (2.2). If $\phi_i(0) = b_i$, then $f_i(\psi) \leq 0$ for every ψ satisfying $\hat{a} \leq \psi \leq \phi$ and $\psi_i(0) = \phi_i(0) = b_i$, by (2.2) again, so $H_i(\phi) \leq 0$ by the definition. Similar arguments apply to h and therefore h and H satisfy (2.2). Now let $\phi, \psi \in \Sigma$ satisfy $\phi \leq \psi$ and $\phi_i(0) = \psi_i(0)$ for some i. Obviously, $H_i(\phi) \leq H_i(\psi)$ since the supremum defining the former is over a subset of the set over which the supremum defining the latter is taken. Again, a similar argument applies to h and hence both functions satisfy (Q).

Next, to show that H satisfies (2.3), we suppose for convenience that $a = 0$. Fix $\eta, \phi \in \Sigma$ and i, with $1 \leq i \leq n$ and let ψ^m satisfy $\hat{0} \leq \psi^m \leq \phi$ and $\psi_i^m(0) = \phi_i(0)$ for $m = 1, 2, \ldots$ such that $H_i(\phi) = \lim_{m \to \infty} f_i(\psi^m)$. For each $m \geq 1$ and each index j, let $\bar{\psi}_j^m(s) = \max\{0, \psi_j^m(s) + \eta_j(s) - \phi_j(s)\}$. Note that $\bar{\psi}^m$ is continuous, $\bar{\psi}_i^m(0) = \eta_i(0)$ and $\hat{0} \leq \bar{\psi}^m \leq \eta$, for $m \geq 1$. Hence,

$$\begin{aligned} H_i(\phi) - H_i(\eta) &= \lim_{m \to \infty} f_i(\psi^m) - H_i(\eta) \leq \limsup_{m \to \infty}[f_i(\psi^m) - f_i(\bar{\psi}^m)] \\ &\leq \limsup_{m \to \infty} L|\psi^m - \bar{\psi}^m|. \end{aligned}$$

It is easy to see from the definition of $\bar{\psi}^m$ that $|\bar{\psi}_j^m(s) - \psi_j^m(s)| \leq |\eta_j(s) - \phi_j(s)|$ for each j and all $s \in [-r, 0]$. Therefore, $H_i(\phi) - H_i(\eta) \leq L|\eta - \phi|$, and by symmetry $|H_i(\phi) - H_i(\psi)| \leq L|\eta - \phi|$. Since we use the maximum norm and i was arbitrary, H satisfies the Lipschitz condition. A similar argument applies to h.

Finally, if f satisfies (Q) and if $\phi \leq \psi \leq \hat{b}$ and $\psi_i(0) = \phi_i(0)$, then $f_i(\phi) \leq f_i(\psi)$. Therefore, $h_i(\phi) = f_i(\phi)$ for $\phi \in \Sigma$ and each i. A similar argument shows that $f = H$. □

The quasimonotone comparison systems related to (0.1) on Σ are given by
$$y'(t) = h(y_t) \tag{2.6}$$
and
$$z'(t) = H(z_t). \tag{2.7}$$
According to Proposition 2.3, Σ is positively invariant for (2.6) and (2.7) so for each $\phi \in \Sigma$, (2.6) and (2.7) have unique solutions $y(t,\phi)$ and $z(t,\phi)$ defined for all $t \geq 0$ and satisfying $a \leq y(t,\phi), z(t,\phi) \leq b$ for all $t \geq 0$. Actually, since h and H satisfy (Q), we have the following important relationship between solutions of (2.6), (2.7) and (0.1).

COROLLARY 2.4. *If $\hat{a} \leq \xi \leq \phi \leq \psi \leq \hat{b}$, then*
$$a \leq y(t,\xi) \leq x(t,\phi) \leq z(t,\psi) \leq b,$$
for all $t \geq 0$ where $x(t,\phi)$ is the solution of (0.1) satisfying $x_0(\phi) = \phi$. Furthermore, if $\xi = \hat{a}$ and $\psi = \hat{b}$, then
$$y(t,\hat{a}) \to c, \quad z(t,\hat{b}) \to d \quad t \to \infty$$
where $a \leq c \leq d \leq b$, $h(\hat{c}) = H(\hat{d}) = 0$, and $y(t,\hat{a})$ is nondecreasing and $z(t,\hat{b})$ is nonincreasing. Finally,
$$\Sigma' = [\hat{c}, \hat{d}]$$
is positively invariant for (0.1).

PROOF. The first assertion follows immediately from Theorem 1.1 and Proposition 2.3. Since h and H satisfy (2.2), $h(\hat{a}) \geq 0$ and $H(\hat{b}) \leq 0$ so $y(t,\hat{a})$ and $z(t,\hat{b})$ are monotone in t by Corollary 2.2. The existence of c, d follow from the positive invariance of Σ under (2.6) and (2.7) and by the first assertion of the Corollary. The positive invariance of Σ' follows from Remark 2.1, Proposition 2.3 and the fact that $h(\hat{c}) = H(\hat{d}) = 0$. For if $\phi \in \Sigma'$ and $\phi_i(0) = c_i$, then $0 = h_i(\hat{c}) \leq h_i(\phi) \leq f_i(\phi)$ where we have used the quasimonotone property of h. □

The important implication of Corollary 2.4 is that we can replace Σ by Σ' and apply the same arguments as above to obtain new comparison functions h' and H' for f relative to Σ'. Clearly, this procedure can be iterated and, under suitable hypotheses, may force solutions of (0.1) to converge. We give such conditions in the next result. The order interval Σ is said to be a *contracting rectangle* for (0.1) provided that the strict inequality $f_i(\phi) > 0$ ($f_i(\phi) < 0$) holds whenever $\phi \in \Sigma$ and $\phi_i(0) = a_i$ ($\phi_i(0) = b_i$) for some i.

THEOREM 2.5. *Suppose there exists $e \in \mathbb{R}^n$ such that $\hat{e} \in \Sigma$ and $f(\hat{e}) = 0$. Assume there exists a one-parameter family of order intervals given by $\Sigma(s) = [\hat{a}(s), \hat{b}(s)]$, $0 \leq s \leq 1$, such that for $0 \leq s_1 < s_2 \leq 1$*
$$a = a(0) \leq a(s_1) < a(s_2) \leq a(1) = e = b(1) \leq b(s_2) < b(s_1) \leq b(0) = b,$$
where $a(s)$ and $b(s)$ are continuous functions of s. Assume that $\Sigma(s)$ is a contracting rectangle for (0.1) for each $s \in [0,1)$. If $\phi \in \Sigma$, then
$$x(t,\phi) \to e, \quad t \to \infty.$$

PROOF. If $\phi \in \Sigma$, then $\omega(\phi)$ is a compact invariant subset of Σ so the set $\{\psi(s) : \psi \in \omega(\phi), s \in [-r, 0]\}$ is a compact subset of the order interval $[a, b] \in \mathbb{R}^n$. This and the continuity of $a(s)$ and $b(s)$ is easily seen to imply the existence of a maximal s with the property that $\omega(\phi) \subset \Sigma(s)$. We label this value of s as s_0. If $s_0 = 1$, then we are done so we assume that $s_0 < 1$. By the maximality of s_0 and the invariance of the omega limit set, there must exist a $\psi \in \omega(\phi)$ and an index i such that $\psi_i(0) = a_i(s_0)$ or $\psi_i(0) = b_i(s_0)$. Suppose the former holds as the argument is similar if the latter holds. Again, by the invariance of $\omega(\phi)$, there exists $\eta \in \omega(\phi)$ such that $x_1(\eta) = \psi$, where $x(t, \eta)$ is the solution of (0.1) satisfying $x_0 = \eta$. Therefore,

$$x'_i(0, \psi) = x'_i(1, \eta) = f_i(\psi) > 0$$

since $\Sigma(s_0)$ is a contracting rectangle. But $x_i(1, \eta) = a_i(s_0)$ so it follows that $x_i(t, \eta) < a_i(s_0)$ for $t < 1$ sufficiently near to $t = 1$. Since $x_t(\eta) \in \omega(\phi)$, this contradicts that $\omega(\phi) \subset \Sigma(s_0)$ and proves the theorem. $\square$

3. Eventual Strong Monotonicity

The central task of this chapter is to set up the framework to apply the results of Chapters 1 and 2 to (0.1). For this, we need to establish sufficient conditions for (0.1) to generate a strongly monotone or strongly order preserving semiflow. However, the framework of the previous sections must be modified in order to give sufficient conditions for strong monotonicity. The following example illustrates the difficulty with our present set up. Consider the system

$$\begin{aligned} x'_1(t) &= -x_1(t) + x_2(t - 1/2) \\ x'_2(t) &= x_1(t - 1) - x_2(t). \end{aligned} \tag{3.1}$$

Observe that (K) holds for (3.1). For initial data, take $\phi = (\phi_1, \phi_2) \in C$ ($r = 1$) where $\phi_1 = 0$ and $\phi_2(\theta) > 0$ for $\theta \in (-1, -2/3)$ and $\phi_2(\theta) = 0$ elsewhere in $[-1, 0]$. This initial value problem can be readily integrated by the method of steps of length $1/2$ and one sees that $x(t) = 0$ for all $t \geq -2/3$. In the language of semiflows, $\phi > 0$ yet $\Phi_t(\phi) = \Phi_t(0) = 0$ for all $t \geq 1/3$. Therefore, the hypotheses of Theorem 1.1 are not sufficient to guarantee that Φ_t is strictly order preserving, that is, $\phi < \psi$ need not imply that $\Phi_t(\phi) < \Phi_t(\psi)$ and Φ_t need not be one-to-one. A little thought will convince the reader that the "problem" is the poor choice of state space for the system, namely, $C([-1, 0], \mathbb{R}^2)$. The remedy is clear; we cannot ignore the fact that different components of the system may require initial data specified on different intervals. We will see that the problem disappears if we choose the state space for the system to be $X = C([-1, 0], \mathbb{R}) \times C([-1/2, 0], \mathbb{R})$.

If X is taken as the state space and initial data $\phi = (\phi_1, \phi_2) \in X$ satisfying $\phi > 0$ (the partial order is, hopefully, obvious but will be formalized later). Then either $\phi_1(\theta) > 0$ for some $\theta \in [-1, 0]$ or $\phi_2(\theta) > 0$ for some $\theta \in [-1/2, 0]$ or both. The method of steps leads to the conclusion that $x(t, \phi) \gg 0$ for $t > 3/2$, and possibly before this. The worst case is when $\phi_2 \equiv 0$, $\phi_1 \geq 0$ and $\phi_1(\theta) > 0$ if and only if $\theta \in (-\epsilon, -\epsilon/2)$ for some small $\epsilon > 0$. For then $x_2(t) = 0$ on $[-1/2, 1 - \epsilon]$ and $x_2(t) > 0$ for $t > 1 - \epsilon$ and $x_1(t) = 0$ on $[-\epsilon/2, 3/2 - \epsilon]$ and $x_1(t) > 0$ for $t > 3/2 - \epsilon$.

Summarizing, if X is the state space for (3.1), then whenever $\phi > 0$ it follows that $\Phi_t(\phi) = x_t(\phi) \gg 0$ for $t > 5/2$, where $x_t(\phi)$ is defined below. Therefore, Φ is eventually strongly monotone, and, in particular, strongly order preserving.

Motivated by this example, let $r = (r_1, r_2, \ldots, r_n) \in \mathbb{R}^n_+$, $R = \max r_i$ and define
$$C_r = \prod_{i=1}^{n} C([-r_i, 0], \mathbb{R}).$$
We write $\phi = (\phi_1, \phi_2, \ldots, \phi_n)$ for a generic point of C_r. C_r is a Banach space with the norm $|\phi| = \sum |\phi_i|$. Let C_r^+ denote the cone of nonnegative functions in C_r and $\leq, <, \ll$ be the corresponding order relations.

If $x_i(t)$ is defined on $[-r_i, \sigma)$, $1 \leq i \leq n$, $\sigma > 0$ then we may redefine $x_t \in C_r$ as $x_t = (x_t^1, x_t^2, \ldots, x_t^n)$ where $x_t^i(\theta) = x_i(t+\theta)$ for $\theta \in [-r_i, 0]$. Notice that now, the subscript signifying a particular component will be raised to a superscript when using the subscript "t" to denote a function. Hereafter, (0.1) and (1.3) will be considered where $f, L(t) : C_r \to \mathbb{R}^n$ and x_t is to be interpreted as described here. There is no change in the basic theory under these circumstances and the results of the previous sections hold with the obvious modifications.

We begin by considering the nonautonomous linear system (1.3). Our goal is to give sufficient conditions for $x_t(\phi) \gg 0$ to hold whenever $\phi > 0$ holds in C_r. The representation (1.4) of Lemma 1.2 must be modified slightly. For $i = 1, 2, \ldots, n$, we have

$$L_i(t)\phi = a_i(t)\phi_i(0) + \sum_{j=1}^{n} \int_{-r_j}^{0} \phi_j(\theta) d_\theta \eta_{ij}(t, \theta) = a_i(t)\phi_i(0) + \bar{L}_i(t)\phi, \quad (3.2)$$

where η_{ij} is a positive Borel measure on $[-r_j, 0]$ and $\bar{L}(t)\phi \geq 0$ whenever $\phi \geq 0$.

By Lemma 1.3, (K) implies that if a component of a solution of (1.3) gets turned on, then it stays on. What we need now is a condition which implies that some component actually gets turned on if the initial condition is nontrivial. The desired condition is as follows.

(R) For each j for which $r_j > 0$, there exists i such that for all t, $\eta_{ij}(t)([-r_j, -r_j + \epsilon)) > 0$ for all small $\epsilon > 0$.

(R) ensures that we have correctly chosen the r_j.

LEMMA 3.1. *Let* (K) *and* (R) *hold. If* $\phi > 0$ *and* $t_0 \in \mathbb{R}$, *then there exists* i *and* $t \in [t_0, t_0 + R)$ *such that* $x_i(t, t_0, \phi) > 0$.

PROOF. If $\phi(0) > 0$ then the conclusion follows from Lemma 1.3 for $t = t_0$. We may assume, therefore, that $r_i > 0$ for some i and that $\phi(0) = 0$. By assumption, there exists j such that $r_j > 0$ and $\theta_0 \in (-r_j, 0)$ such that $\phi_j(\theta_0) > 0$. Let i be as in (R) corresponding to j and set $t_1 = t_0 + r_j + \theta_0$. If $x_i(t_1) > 0$, then the proof is complete since $t_1 > t_0$. If $x_i(t_1) = 0$ then

$$x_i'(t_1) = L_i(t_1) x_{t_1}$$
$$\geq \int_{-r_j}^{0} x_i(t_1 + \theta) d_\theta \eta_{ij}(t_1, \theta)$$
$$\geq \int_{-r_j}^{-r_j - \theta_0} \phi_j(\theta + r_j - \theta_0) d_\theta \eta_{ij}(t_1, \theta)$$
$$> 0$$

where we have used (R) and $\phi_j(\theta_0) > 0$ to conclude the final inequality. $\square$

(R) ensures that some component of a nontrivial, nonnegative solution of (1.3) gets turned on and (K) ensures that it stays on. We now require a condition which will ensure that once one component gets turned on then all components eventually get turned on. This is a kind of irreducibility assumption.

(I) The matrix $A(L)(t)$ defined by
$$A(L)(t) = \text{col}(L(t)\hat{e}_1, L(t)\hat{e}_2, \ldots, L(t)\hat{e}_n)$$
is irreducible.

If (K) holds then (1.5) implies that
$$A(L)_{ij}(t) = a_i(t)\delta_{ij} + \eta_{ij}(t)([-r_j, 0]) \tag{3.3}$$
where δ_{ij} is the Kronecker delta satisfying $\delta_{ij} = 1$ if $i = j$ and $\delta_{ij} = 0$ otherwise. (I) holds if and only if $B(t)$ is irreducible where $B_{ij}(t) = \bar{L}_i(t)\hat{e}_j = \eta_{ij}(t)([-r_j, 0])$.

LEMMA 3.2. *Let* (K), (R) *and* (I) *hold. If* $\phi > 0$ *and* t_0 *are given, then* $x(t, t_0, \phi) \gg 0$ *for* $t \geq t_0 + nR$.

PROOF. Let $t_j = t_0 + jR$, $j = 0, 1, \ldots$. By Lemma 1.3 and Lemma 3.1, there exists i such that $x_i(t) > 0$ for $t \geq t_1$. It follows that there exists $\alpha(t) > 0$ for $t \geq t_2$ such that $x_t \geq \alpha(t)\hat{e}_i$ for $t \geq t_2$. Hence, $\bar{L}(t)x_t \geq \alpha(t)\bar{L}(t)\hat{e}_i$ for $t \geq t_2$. Since $B(t)$ is irreducible there exists $j \neq i$ such that $B_{ji}(t_2) > 0$. Either $x_j(t_2) > 0$ or $x_j(t_2) = 0$ and
$$x'_j(t_2) = \bar{L}_j(t_2)x_{t_2} \geq \alpha(t_2)\bar{L}_j(t_2)\hat{e}_i = \alpha(t_2)B_{ji}(t_2) > 0.$$
But $x_j(t_2) = 0$ and $x'_j(t_2) > 0$ contradicts that $x_j(t) \geq 0$ for $t \geq t_0$. Hence, $x_j(t_2) > 0$ and by Lemma 1.3, $x_j(t) > 0$ for $t \geq t_2$.

It follows that there exists $\beta(t) > 0$ for $t \geq t_3$ such that $x_t \geq \alpha(t)\hat{e}_i + \beta(t)\hat{e}_j$ for $t \geq t_3$. Thus,
$$\bar{L}(t)x_t \geq \alpha(t)\bar{L}(t)\hat{e}_i + \beta(t)\bar{L}(t)\hat{e}_j = B(t)(\alpha(t)e_i + \beta(t)e_j).$$
Since $B(t_3)$ is irreducible, there exists $k \neq i, j$ such that $B_{kl}(t_3) > 0$ for either $l = i$ or $l = j$ or both. If $x_k(t_3) = 0$ then, assuming $l = i$
$$x'_k(t_3) = \bar{L}_k(t_3)x_{t_3} \geq B(t_3)(\alpha(t_3)e_i + \beta(t_3)e_j)) \cdot e_k \geq B_{ki}(t_3)\alpha(t_3) > 0.$$
A similar argument shows that $x'_k(t_3) > 0$ if $l = j$. Again, this contradicts that $x_k(t) \geq 0$ for $t \geq t_0$. Therefore, $x_k(t_3) > 0$ and by Lemma 1.3, $x_k(t) > 0$ for $t \geq t_3$. Continuing in this manner, we obtain that $x(t) \gg 0$ for $t \geq t_n$. $\square$

In example (3.1), $L(t)\phi$ is given by
$$L(t)\phi = \text{col}(-\phi_1(0) + \phi_2(-1/2), \phi_1(-1) - \phi_2(0)).$$
$a_i(t) = -1$ and $\eta_{ii} = 0$ for $i = 1, 2$; $\eta_{12} = \delta_{-1/2}$ and $\eta_{21} = \delta_{-1}$ where δ_s is the Dirac measure satisfying $\delta_s(E) = 1$ if $s \in E$ and $\delta(E) = 0$ otherwise. It is obvious that (R) holds if $r = (1, 1/2)$. Since
$$A(L)(t) = \begin{pmatrix} -1 & 1 \\ 1 & -1 \end{pmatrix}$$

is irreducible, it follows that (I) holds as well.

Now let D be an open subset of C_r, $f : D \to \mathbb{R}^n$ be continuously differentiable and consider
$$x'(t) = f(x_t). \tag{3.4}$$
As a result of the smoothness assumption on f, it is Lipschitz on each compact subset of D and consequently initial value problems associated with (3.4) have unique solutions. Hereafter, we assume that for each $\phi \in D$, the solution $x(t, \phi)$ of (3.4) satisfying $x_0 = \phi$ is defined for all $t \geq 0$. In this case, (3.4) defines a semiflow Φ on D, defined by
$$\Phi_t(\phi) = x_t(\phi), \quad \phi \in D.$$

We say that (3.4) is *cooperative* if D is order convex and if $df(\phi)$ satisfies (K) for each $\phi \in D$. In that case, the derivative $df(\phi)$ can be represented as in (3.2) where $a_i = a_i(\phi)$ and $\eta_{ij} = \eta_{ij}(\phi)$ are continuous functions of $\phi \in D$. (3.4) is said to be *cooperative and irreducible* if it is cooperative and the following hold:
(1) $df(\phi)$ satisfies (I) for each $\phi \in D$.
(2) For every j for which $r_j > 0$, there exists i such that for all $\phi \in D$
$$\eta_{ij}(\phi)([-r_j, -r_j + \epsilon)) > 0$$
for all small $\epsilon > 0$.

LEMMA 3.3. *If (3.4) is cooperative in D, then f satisfies* (Q).

PROOF. Suppose that $\phi, \psi \in D$ satisfy $\phi \leq \psi$ and $\phi_i(0) = \psi_i(0)$. Since D is order convex and f is continuously differentiable, it follows that
$$f_i(\psi) - f_i(\phi) = \int_0^1 df_i(s\psi + (1-s)\phi)(\psi - \phi)ds \geq 0,$$
where the last inequality holds because by (K), the integrand is nonnegative. □

THEOREM 3.4. *Let f be cooperative and irreducible in D. If $\phi, \psi \in D$ satisfy $\phi < \psi$ then*
$$x(t, \phi) \ll x(t, \psi) \quad t \geq nR.$$

PROOF. By Theorem 1.1 and Lemma 3.3, $x(t, \phi) \leq x(t, \psi)$ for $t \geq 0$. Since $x(t, \phi)$ is continuously differentiable in its second variable (Hale(1977)), we have
$$x(t, \psi) - x(t, \phi) = \int_0^1 d_\phi x(t, s\psi + (1-s)\phi)(\psi - \phi)ds.$$
If $\chi \in D$ and $\beta \in C_r$, then $d_\phi x(t, \chi)\beta = y(t, \beta)$ satisfies the linear variational equation
$$y'(t) = df(x_t(\chi))y_t, \quad y_0 = \beta.$$
See Hale(1977) Chapter 2, Theorem 4.1. If $L(t) = df(x_t(\chi))$, then it is apparent that $L(t)$ satisfies (K),(I) and (R) on $t \geq 0$. By Lemma 3.2, if $\beta \in C_r$ satisfies $\beta > 0$, then $y(t, \beta) \gg 0$ for $t \geq nR$. Since $\psi - \phi > 0$, it follows that the integrand above is positive for $t \geq nR$. This proves the assertion. □

An immediate corollary of Theorem 3.4 follows. It is the main result of this section.

COROLLARY 3.5. *If (3.4) is cooperative and irreducible in D and $\phi, \psi \in D$ satisfy $\phi < \psi$ then*
$$x_t(\phi) \ll x_t(\psi), \quad t \geq (n+1)R.$$
In particular, the semiflow Φ is eventually strongly monotone.

As an example, consider (1.2) where $n = 1$, that is, where (1.2) is a scalar equation. We suppose that $g(x, y)$ is continuously differentiable on $\mathbb{R}^2$. (K) holds if
$$g_y(x,y) = \frac{\partial g}{\partial y} \geq 0.$$
For $C = C_r$ where $r > 0$ is a scalar, define $f : C \to \mathbb{R}$ by $f(\phi) = g(\phi(0), \phi(-r))$. Then
$$df(\phi)\psi = g_x(\phi(0), \phi(-r))\psi(0) + g_y(\phi(0), \phi(-r))\psi(-r)$$
$$= a(\phi)\psi(0) + \int_{-r}^{0} \psi(\theta) d_\theta \eta(\phi, \theta),$$
where
$$a(\phi) = g_x(\phi(0), \phi(-r))$$
and
$$\eta(\phi) = g_y(\phi(0), \phi(-r))\delta_{-r}.$$
Consequently,
$$\eta(\phi)([-r, -r+\epsilon)) = g_y(\phi(0), \phi(-r)).$$
Therefore, (R) holds if and only if
$$g_y(x,y) > 0 \tag{3.5}$$
for all $(x, y) \in \mathbb{R}^2$. (I) is trivially satisfied for a scalar equation. Therefore, Corollary 3.5 applies to (1.2) if (3.5) holds.

In the general case of a system, as noted in section 1, (1.2) is cooperative provided that the Jacobian matrices satisfy $\frac{\partial g}{\partial x}$ is quasipositive and $\frac{\partial g}{\partial y} \geq 0$. The reader may check the following calculations
$$\eta_{ij}(\phi)([-r, -r+\epsilon)) = \frac{\partial g_i}{\partial y_j}$$
and
$$A(L(\phi)) = \frac{\partial g}{\partial x} + \frac{\partial g}{\partial y},$$
where the argument of each partial derivative is $(\phi(0), \phi(-r))$. By definition, (I) holds if and only if $A(L(\phi))$ is irreducible for every $\phi \in D$. (R) holds if for each j and each (x, y) there exists i such that $\frac{\partial g_i}{\partial y_j}(x, y) > 0$.

4. Generic Convergence for Cooperative and Irreducible Systems

The aim of this section is to apply the main result of Chapter 2 to cooperative and irreducible functional differential equations to conclude that the generic solution converges to equilibrium. Consider the system
$$x'(t) = f(x_t) \tag{4.1}$$
where $f : D \to \mathbb{R}^n$ is continuously differentiable on D and $D \subset C_r$. In order to satisfy the compactness requirement (C) of Chapters 1 and 2, we introduce the following assumption.

(T) f maps bounded subsets of D to bounded subsets of $\mathbb{R}^n$. For each $\phi \in D$, $x(t,\phi)$ is defined and bounded for $t \geq 0$. Moreover, for each compact subset A of D, there exists a closed and bounded subset $B = B(A)$ of D such that for each $\phi \in A$, $x_t(\phi) \in B$ for all large t.

The main result of this chapter is the following.

THEOREM 4.1. *Let (4.1) be cooperative and irreducible in D and suppose that (T) holds. Then the set of convergent points in D contains an open and dense subset.*

PROOF. We must verify the hypotheses (M),(D),(S) and (C) of Theorem 2.4.7. Corollary 3.5 implies that (M) holds with $\tau = (n+1)R$. As noted in the proof of Theorem 3.4, $x(t,\phi)$ is continuously differentiable in ϕ for fixed $t \geq 0$ and $d_\phi x(t,\phi)\beta = y(t,\beta)$, the unique solution of the linear variational equation

$$y'(t) = df(x_t(\phi))y_t, \quad y_0 = \beta. \tag{4.2}$$

It is not difficult to see that $\Phi'_\tau(\phi)\beta = y_\tau(\beta)$ and therefore, Φ_τ is continuously differentiable as required for (D) to hold. See also Hale(1988), Thm.4.1.1. Moreover, $\Phi'_\tau(\phi)$ is strongly positive since, by Lemma 3.2, $y_\tau(\beta) \gg 0$ whenever $\beta > 0$. Thus (S) holds. We now show that (C) holds as a consequence of (T). Since $x_t(\phi)$ is bounded for $t \geq 0$ and f maps bounded subsets of D to bounded subsets of $\mathbb{R}^n$, it follows that $x'(t,\phi)$ is uniformly bounded on $t \geq 0$. By (T) with $A = \{\phi\}$, there is a closed and bounded set $B \subset D$ such that $x_t(\phi) \in B$ for all large t. Consequently, by the Ascoli-Arzela Theorem, the orbit $\{x_t(\phi) : t \geq 0\}$ has compact closure in D. Now if A is a compact subset of D and B' is the closure in D of the ϵ-neighborhood of $B(A)$ in D, then by (T) there exists $t_0 > 0$ such that $x_t(\phi) \in B'$ for all $\phi \in A$ and for all $t \geq t_0$. This is easily seen to imply that (C) holds since, if ϕ_n approximates ϕ from below or from above in D then we take $A = \{\phi_n\} \cup \{\phi\}$. □

As a simple application of Theorem 4.1, consider the scalar equation

$$x'(t) = -x(t) + h(x(t-r)) \tag{4.3}$$

where $r > 0$ and $h : \mathbb{R} \to \mathbb{R}$ is continuously differentiable and satisfies

$$h'(x) > 0, \quad x \in \mathbb{R} \tag{4.4}$$

and

$$\limsup_{|x|\to\infty} \frac{h(x)}{x} < 1. \tag{4.5}$$

The set of equilibria for (4.3) is given by

$$E = \{\hat{x} : x \in \mathbb{R} \text{ and } x = h(x)\}.$$

PROPOSITION 4.2. *There is an open and dense set of initial conditions in C_r corresponding to solutions of (4.3) which converge to an equilibrium. If E consists of a single element, then all solutions converge to it. If E consists of two elements, then all solutions converge to one of these.*

PROOF. (4.4) guarantees that (4.3) is cooperative and irreducible in C_r. By (4.5), $h(M)/M < 1$ and $h(-M)/(-M) < 1$, for all large M. Therefore, if $\phi = \hat{M}$, then $-\phi(0) + h(\phi(-r)) = -M + h(M) < 0$; similarly, if $\phi = -\hat{M}$, then $-\phi(0) + h(\phi(-r)) = M + h(-M) > 0$. It follows from Corollary 2.2 that $x(t, \hat{M})$ is monotonically decreasing and $x(t, -\hat{M})$ is monotonically increasing. Since $-\hat{M} < \hat{M}$, monotonicity implies that $-\hat{M} < x_t(-\hat{M}) \ll x_t(\hat{M}) < \hat{M}$ for $t \geq 0$. Thus, both orbits are bounded and since the right side of (4.3) is bounded on bounded subsets of C_r, both orbits have compact closure in C_r. By Corollary 2.2, $x_t(-\hat{M})$ and $x_t(\hat{M})$ converge to an element of E as $t \to \infty$. But E is bounded by (4.5) and therefore there is a closed and bounded set B, containing E in its interior, such that $x_t(-\hat{M}), x_t(\hat{M}) \in B$ for all large t. Furthermore, B can be taken to be $B = [-\hat{L}, \hat{L}]$ for a sufficiently large $L > 0$. It follows that $x_t(\pm \hat{M}) \in D$ for all large t and that this holds for all large M as well. Given $\phi \in C_r$ there is an $M > 0$ such that $-\hat{M} < \phi < \hat{M}$ and monotonicity then implies that $x_t(-\hat{M}) \leq x_t(\phi) \leq x_t(\hat{M})$ for $t \geq 0$. Consequently, $x_t(\phi) \in B$ for all large t. Thus (T) holds. The first assertion now follows from Theorem 4.1. The second assertion follows from Theorem 2.3.1 and the final assertion follows from Theorem 2.3.2. □

Actually, because the set E in example (4.3) is a totally ordered set in C_r, we could as well have applied Theorem 1.4.3 and Remark 4.2 of Chapter 1 to conclude Proposition 4.2.

5. Stability of Equilibria

In this section we are concerned with the stability of an equilibrium of the system
$$x'(t) = f(x_t) \tag{5.1}$$
where f is continuously differentiable and cooperative in a domain D. Suppose that $\hat{v}$ is an equilibrium of (5.1), that is, $v \in \mathbb{R}^n$ is such that $\hat{v} \in D$ and $f(\hat{v}) = 0$. The linear variational system corresponding to $\hat{v}$ is
$$y'(t) = Ly_t, \quad L = df(\hat{v}). \tag{5.2}$$
The stability of the trivial solution of (5.2) is determined by the characteristic equation obtained by seeking solutions of (5.2) of the form $y(t) = e^{\lambda t} u$ where $u \in \mathbb{R}^n$. If L is represented as in (3.2), then λ must be a root of
$$\text{Det}\Delta(\lambda) = 0 \tag{5.3}$$
where
$$\Delta(\lambda) = \lambda I - A(\lambda)$$
and
$$A(\lambda)_{ij} = a_i \delta_{ij} + \int_{-r_j}^{0} e^{\lambda \theta} d\eta_{ij}(\theta). \tag{5.4}$$
The *stability modulus* of L is defined as
$$s(L) = \max\{\Re \lambda : \text{Det}\Delta(\lambda) = 0\}.$$
It is well-defined since for any $\beta \in \mathbb{R}$, there are at most a finite number of roots of (5.3) satisfying $\Re \lambda \geq \beta$. $\hat{v}$ is asymptotically stable if $s(L) < 0$ and unstable if $s(L) > 0$.

The linearized stability of an equilibrium of a cooperative and irreducible system can be determined by considering only the real roots of the characteristic equation (5.3).

THEOREM 5.1. *Let L satisfy (K),(R) and (I). Then $s(L)$ is a root of (5.3) of algebraic multiplicity one. Furthermore, there is a solution $y(t) = e^{s(L)t}u$ of (5.2) where $u \gg 0$. If λ is any other root of (5.3) then $\Re\lambda < s(L)$.*

PROOF. We begin by considering $A(\lambda)$ for real values of λ. First observe from (5.4) that $A(\lambda)$ is quasipositive. It is also irreducible. In fact, $A(0) = A(L(\hat{v}))$ as is easily seen from (3.3) and $A(\lambda)_{ij} = 0$ if and only if $A(0)_{ij} = 0$. Clearly, $s(A(\lambda))$ is a continuous function of λ. The map $\lambda \to A(\lambda)$ is nonincreasing from (5.4) by monotonicity of the exponential function in λ. It follows from Corollary 4.3.2 (3) that $s(A(\lambda_1)) \geq s(A(\lambda_2))$ where $\lambda_1 \leq \lambda_2$. This implies that the equation $s(A(\lambda)) = \lambda$ has a solution, $\lambda = \lambda_0$. Furthermore, this solution is unique as it is a root of $\lambda - s(A(\lambda)) = 0$ and $\lambda - s(A(\lambda))$ is strictly increasing in λ.

We will show that $s(L) = \lambda_0$ but first we show that it is the largest real root of (5.3). Since $s(A(\lambda_0)) = \lambda_0$ is an eigenvalue of the quasipositive matrix $A(\lambda_0)$, it follows that

$$\text{Det}(\lambda_0 I - A(\lambda_0)) = 0$$

and therefore, λ_0 is a root of (5.3). According to Corollary 4.3.2 to the Perron-Frobenius Theorem, there is an eigenvector $u \gg 0$ of $A(\lambda_0)$ corresponding to the eigenvalue λ_0. An easy computation shows that $y(t) = e^{\lambda_0 t}u$ is a solution of (5.2).

Let λ be any real root of (5.3). Then λ is an eigenvalue of $A(\lambda)$ so, by the definition of the stability modulus, $\lambda \leq s(A(\lambda))$, or $\lambda - s(A(\lambda)) \leq 0$. It follows that $\lambda \leq \lambda_0$. Thus λ_0 is the largest real root of (5.3).

In the remainder of the proof, a different approach is taken, exploiting the fact that the linear variational equation (5.2) generates a strongly continuous semigroup of positive operators $\{T(t)\}_{t \geq 0}$, defined by

$$T(t)\beta = y_t(\beta), \quad \beta \in C_r.$$

By the Ascoli-Arzela Theorem and a Gronwall estimate, it is easy to see that $T(t)$ is compact for $t \geq R$. The standard theory of semigroups (see e.g. Hale(1977)) implies that if M is the infinitesimal generator of the semigroup $\{T(t)\}$, defined on its domain $D(M)$ by $M\phi = \lim_{t \to 0+} t^{-1}(T(t)\phi - \phi)$, then the *point spectrum* of $T(t)$, $P\sigma(T(t)) = \{\lambda : N(\lambda I - T(t)) \neq 0\}$, satisfies $P\sigma(T(t)) \setminus \{0\} = \exp(tP\sigma(M))$. There is no need to identify the generator M of $T(t)$ here but see Hale (1977). Since $T(t)$ is compact for $t \geq R$, it follows that $\sigma(T(t)) = P\sigma(T(t)) \cup \{0\}$ for these t. Now Hale(1977) shows that $P\sigma(M) = \sigma(M)$ consists of the roots of (5.3). Therefore,

$$\sigma(T(t)) = \{0\} \cup \{\exp(\lambda t) : \lambda \text{ is a root of } (5.3)\}.$$

Consequently, if $\rho(T(t))$ is the spectral radius of $T(t)$, then

$$\rho(T(t)) = \exp(s(L)t).$$

As observed in the proof of Theorem 4.1, $T(t)$ is strongly positive for $t \geq (n+1)R$ by Lemma 3.2. Since $y(t) = e^{\lambda_0 t}u$ is a solution of (5.2), we conclude that

$$T(t)(\psi) = e^{\lambda_0 t}\psi$$

where $\psi_i(\theta) = u_i \exp(\lambda_0 \theta)$, $-r_j \leq \theta \leq 0$. But $\psi \gg 0$ so by the uniqueness of the positive eigenvector of $T(t)$, guaranteed by the Krein-Rutman Theorem (Theorem 2.4.1), $\rho(T(t)) = e^{\lambda_0 t}$. Therefore, $s(L) = \lambda_0$. Also by the Krein-Rutman Theorem, $e^{s(L)t}$ has algebraic multiplicity one as an eigenvalue of $T(t)$. From this and Lemma 4.1, Chapter 7 of Hale(1977), it follows that $s(L)$ has algebraic multiplicity one as a root of (5.3). If λ satisfies (5.3) and $\lambda \neq \lambda_0$, then as noted above, $e^{\lambda t}$ is an eigenvalue of $T(t)$. $e^{\lambda t} \neq e^{\lambda_0 t}$ for some t and therefore, by the Krein-Rutman Theorem again, $|e^{\lambda t}| = e^{\Re \lambda t} < e^{\lambda_0 t}$ or $\Re \lambda < \lambda_0$. □

A cooperative and irreducible system of ordinary differential equations can be associated with (5.1) simply by ignoring any delays which appear in (5.1). This leads to the system

$$x' = F(x), \quad F(x) = f(\hat{x}) \qquad (5.5)$$

Observe that (5.5) has the same equilibria as (5.1). In particular, v is an equilibrium of (5.5). System (5.5) is cooperative and irreducible since by (3.3)

$$DF(x) = df(\hat{x})(\hat{e}_1, \hat{e}_2, \ldots, \hat{e}_n) = A(0).$$

where the i-th column of the matrix in the middle is $df(\hat{x})\hat{e}_i$. Since $A(0)$ is quasipositive and since (I) holds, $DF(x)$ is a quasipositive and irreducible matrix. It is a remarkable fact that v is asymptotically stable (unstable) in the linear approximation for (5.5) if and only if $\hat{v}$ is asymptotically stable (unstable) in the linear approximation.

COROLLARY 5.2. $s(L) < 0$ ($s(L) > 0$) if and only if $s(DF(v)) < 0$ ($s(DF(v)) > 0$).

PROOF. Recall that $s(L) = \lambda_0$, the unique real root of $\lambda - s(A(\lambda)) = 0$. The map $\lambda \to \lambda - s(A(\lambda))$ is an increasing continuous function. Hence $\lambda_0 < 0$ if and only if $0 - s(A(0)) > 0$, that is, if and only if $s(A(0)) < 0$. Similarly, $\lambda_0 > 0$ if and only if $s(A(0)) > 0$. □

According to the Corollary, the linear stability analysis of $\hat{v}$ for (5.1) may as well be carried out by linearizing (5.5) about v.

The system (5.5) has other relationships to (5.1) as well. For instance, the condition for the existence of a monotone trajectory initiating from $\hat{h}$, given in Corollary 2.2, namely, $f(\hat{h}) \geq 0$ or $f(\hat{h}) \leq 0$, is equivalent to $F(h) \geq 0$ or $F(h) \leq 0$, which is precisely the condition for a monotone trajectory of (5.5) initiating from h according to Proposition 3.2.1. This equivalence is exploited in the next example.

6. A Biochemical Control Circuit with Delays

As an example of the application of some of the ideas developed in this chapter, we revisit the biochemical control circuit considered in Chapter 4, section 2. The production of the various proteins is, of course, not instantaneous and it is reasonable to introduce time delays into these terms. If we do so, the result is the system

$$\begin{aligned} x_1'(t) &= g(x_n(t - r_n)) - \alpha_1 x_1(t) \\ x_j'(t) &= x_{j-1}(t - r_{j-1}) - \alpha_j x_j(t), \quad 2 \leq j \leq n \end{aligned} \qquad (6.1)$$

where $r_j \geq 0$, $\alpha_j > 0$, $g : \mathbb{R}_+ \to \mathbb{R}_+$ is continuously differentiable and there exists $M > 0$ such that
$$0 < g(u) < M, \quad g'(u) > 0, \quad u > 0.$$

The appropriate domain for (6.1) is $D = C_r^+$, the cone of nonnegative functions in C_r where $r = (r_1, r_2, \ldots, r_n)$. Theorem 2.1 implies that D is positively invariant.

As noted in the previous section, the equilibria for (6.1) are obtained by ignoring the delays and therefore they are the same as for the system (2.1) treated in Chapter 4. As in that chapter, we denote by $\hat{E}_*$ and $\hat{E}_{**}$ the minimal and maximal equilibria of (6.1). By Corollary 5.2, the stability properties of these equilibria are the same as for the system without delays.

System (6.1) is cooperative and irreducible in D. In fact, for $\phi \in D$ we have
$$df_1(\phi)\psi = -\alpha_1 \psi_1(0) + g'(\phi_n(-r_n))\psi_n(-r_n)$$
and
$$df_j(\phi)\psi = -\alpha_j \psi_j(0) + \psi_{j-1}(-r_{j-1}), \quad 2 \leq j \leq n.$$
(R) follows from
$$\eta_{(j+1)j}([-r_j, -r_j + \epsilon)) = 1, \quad j \neq n, \quad \eta_{1n}([-r_n, -r_n + \epsilon)) = g'(\phi_n(-r_n))$$
for $0 < \epsilon < \min r_i$. The off-diagonal entries of the matrix $A(L(\phi))$ are zero except for the subdiagonal entries and the entry in the first row and last column, all of which are positive. Thus, $A(L(\phi))$ is irreducible and (I) holds.

The next result is almost identical to Proposition 2.1 of Chapter 4. The introduction of time delays has no effect on the long term dynamics of these control systems.

PROPOSITION 6.1. *All orbits of (6.1) are attracted to $[\hat{E}_*, \hat{E}_{**}]$. There is an open and dense subset of D consisting of convergent points. If E is a single equilibrium, then it attracts all orbits. If E consists of two points, then each orbit is attracted to one of these points.*

PROOF. The proof is a carbon copy of the proof of Proposition 4.2.1. If we write $f(x_t)$ for the right side of (6.1), then $f(\hat{0}) \geq 0$ so $x_t(\hat{0})$ is nondecreasing by Corollary 2.2. If $w \in \mathbb{R}^n$ is defined as $w = M(\alpha_1^{-1}, (\alpha_1 \alpha_2)^{-1}, \ldots, (\alpha_1 \alpha_2 \ldots \alpha_n)^{-1})$, then $f(s\hat{w}) < 0$ for all $s \geq 1$. Thus $x_t(\hat{0})$ converges to $\hat{E}_*$ and $x_t(s\hat{w})$ converges to E_{**} as $t \to \infty$. The remainder of the proof is similar to that of Proposition 4.2.1 and Proposition 4.2. $\square$

7. Competition with Time Delays

The comparison results of section 2 are illustrated in this section by applying them to the Lotka-Volterra competition system with time delays in the competition terms. The equation for the density of population u_i, $1 \leq i \leq n$, is given by

$$u_i'(t) = b_i u_i(t) \Big[1 - c_{ii}\Big(a_i u_i(t) + (1 - a_i) \int_{-r}^0 u_i(t+s)d\nu_{ii}(s)\Big) \\ - \sum_{j \neq i} c_{ij} \int_{-r}^0 u_j(t+s)d\nu_{ij}(s) \Big] \tag{7.1}$$

where
$$c_{ii} > 0,\ b_i > 0,\ c_{ij} \geq 0,\ 0 < a_i < 1,\ r > 0$$
and ν_{ij} are positive Borel measures satisfying
$$\nu_{ij}([-r,0]) = 1,\quad \nu_{ii}(\{0\}) = 0.$$

The first condition on the measures ν_{ij} is simply a normalization ensuring that the total weight of the competition of population u_j on population u_i is measured by the competition coefficient c_{ij}. The self-limitation or crowding term in (7.1),
$$a_i u_i(t) + (1 - a_i) \int_{-r}^{0} u_i(t+s) d\nu_{ii}(s),$$
is explicitly separated into an "instantaneous" part, $a_i u_i(t)$, and a delayed part, $(1-a_i)\int_{-r}^0 u_i(t+s)d\nu_{ii}(s)$. The effect of the second condition on ν_{ii} is simply to ensure that the delayed part has no "instantaneous component". The constant a_i, which satisfies $0 < a_i < 1$, is a measure of the relative strength of the instantaneous intraspecific crowding effect relative to the corresponding delayed effect. If a_i is near zero, then the crowding effect is primarily delayed and if a_i is near one, then it is primarily instantaneous. The methods used here require that $a_i > 0$.

For $R = (R_1, R_2, \ldots, R_n) > 0$, set
$$\Sigma_R = [\hat{0}, \hat{R}].$$
It is shown below that Σ_R is positively invariant for (7.1) for suitable R. We let $f_i(u_t)$ denote the right side of (7.1).

LEMMA 7.1. *If $R_i \geq (c_{ii}a_i)^{-1}$, $1 \leq i \leq n$, then Σ_R is positively invariant for (7.1).*

PROOF. Let $\phi \in \Sigma_R$ where R satisfies the above hypotheses. If $\phi_i(0) = 0$, then $f_i(\phi) = 0$. If $\phi_i(0) = R_i$, then
$$f_i(\phi) \leq b_i R_i[1 - c_{ii} a_i R_i] \leq 0$$
since $R_i \geq (a_i c_{ii})^{-1}$. The result follows from Theorem 2.1 and Remark 2.1. □

We now seek comparison systems corresponding to (7.1). Given an initial condition $\phi \in C_+$ for (7.1), choose R_i satisfying the hypotheses of Lemma 7.1 such that $\phi \in \Sigma_R$. Then the system corresponding to (2.6) for (7.1) on Σ_R is given by
$$y_i'(t) = -b_i y_i(t)[k_i + c_{ii} a_i y_i(t)] \tag{7.2}$$
where $k_i = c_{ii}(1-a_i)R_i + \sum_{j \neq i} c_{ij} R_j - 1$ and the system corresponding to (2.7) is given by
$$z_i'(t) = b_i z_i(t)[1 - c_{ii} a_i z_i(t)]. \tag{7.3}$$
These assertions and their consequences are stated formally below.

PROPOSITION 7.2. *Let $y(t)$ be the solution of (7.2) satisfying $y(0) = \phi(0)$ and $z(t)$ be the solution of (7.3) satisfying $z(0) = \phi(0)$. Then*
$$y(t) \leq u(t, \phi) \leq z(t)$$
for all $t > 0$. In particular, $u_i(t) > 0$ for $t > 0$ if $\phi_i(0) > 0$ and
$$\limsup_{t \to \infty} u_i(t) \leq (c_{ii} a_i)^{-1}.$$

PROOF. The result follows from Corollary 2.4 and the following calculations. If $\phi \in \Sigma_R$, then by (2.4),

$$H_i(\phi) = \sup\Big\{b_i\phi_i(0)\Big[1 - c_{ii}\big(a_i\phi_i(0) + (1-a_i)\int_{-r}^0 \psi_i d\nu_{ii}\big)$$
$$- \sum_{j\neq i} c_{ij}\int_{-r}^0 \psi_j d\nu_{ij}\Big] : 0 \leq \psi \leq \phi, \phi_i(0) = \psi_i(0)\Big\}$$
$$= b_i\phi_i(0)[1 - c_{ii}a_i\phi_i(0)].$$

This computation used the fact that $\nu_{ii}(\{0\}) = 0$. Similarly,

$$h_i(\phi) = \inf\Big\{b_i\phi_i(0)\Big[1 - c_{ii}\big(a_i\phi_i(0) + (1-a_i)\int_{-r}^0 \psi_i d\nu_{ii}\big)$$
$$- \sum_{j\neq i} c_{ij}\int_{-r}^0 \psi_j d\nu_{ij}\Big] : \phi \leq \psi \leq \hat{R}, \phi_i(0) = \psi_i(0)\Big\}$$
$$= b_i\phi_i(0)\Big[1 - c_{ii}\big(a_i\phi_i(0) + (1-a_i)R_i\big) - \sum_{j\neq i} c_{ij}R_j\Big].$$

These calculations show that (7.2) and (7.3) correspond to (2.6) and (2.7). Now we can view (7.2) and (7.3) as delay equations although they are clearly ordinary differential equations. According to Corollary 4.2, the inequality in the proposition holds if we take the initial data $y_0 = \phi$ and $z_0 = \phi$. □

Conditions for the existence of a unique equilibrium representing the coexistence of all n populations are given below. They are essentially that the self-limitation terms (c_{ii}) dominate the interspecific competition terms (c_{ij}). The proof follows Brown(1980).

PROPOSITION 7.3. *If*

$$\sum_{j=1}^n c_{ij}c_{jj}^{-1} < 2, \quad 1 \leq i \leq n,$$

then (7.1) has a unique positive equilibrium $u^* \gg 0$.

PROOF. Let $F(u) = (F_1, F_2, \ldots, F_n)$ be the affine vector field given by $F_i(u) = 1 - \sum_{j=1}^n c_{ij}u_j$. Let $v = (c_{11}^{-1}, c_{22}^{-1}, \ldots, c_{nn}^{-1})$ and let $\Gamma = [0, v + \epsilon 1] \subset \mathbb{R}^n$ where $\epsilon > 0$ and 1 is the vector with all components equal one. We claim that Γ is a contracting rectangle for F if ϵ is sufficiently small. For if $u_i = 0$, then

$$F_i(u) \geq 1 - \sum_{j\neq i} c_{ij}(c_{jj}^{-1} + \epsilon) > 0$$

where the last inequality follows from the hypothesis if ϵ is sufficiently small. If $u_i = v_i + \epsilon$, then

$$F_i(u) \leq 1 - c_{ii}(c_{ii}^{-1} + \epsilon) < 0.$$

It follows that the order interval Γ is positively invariant for the affine system $u' = F(u)$ and by a standard argument using the Brouwer fixed point theorem, there exists an equilibrium for this system in the interior of Γ. See Hale(1980), Theorem 1.8.2. This gives the existence of a positive equilibrium of (7.1).

If there are two distinct positive solutions u^* and v^* of $0 = F(u)$, then $tu^* + (1-t)v^*$ is also a solution for every t and consequently there must be a solution belonging to the boundary of Γ. But this is impossible since Γ is contracting. Therefore, there is exactly one positive solution u^* of $F(u) = 0$. This implies that (7.1) has the unique positive equilibrium u^*. $\square$

Our goal is to show that u^* is globally attracting for (7.1). For this, we require the following condition which is stronger than the hypothesis of Proposition 7.3.

$$\sum_{j=1}^{n} c_{ij}(c_{jj}a_j)^{-1} < 2, \quad 1 \leq i \leq n. \tag{7.4}$$

In addition to requiring that the self-limitation terms dominate the interspecific competition terms, (7.4) requires that the a_i are nearer to one than to zero. That is, the instantaneous self-limitation effect dominates the corresponding delayed effect. In the extreme case that $n = 1$, (7.4) reduces to $a_1 > 1/2$. This condition guarantees that a family of contracting rectangles exists for (7.1).

LEMMA 7.4. *Suppose that* (7.4) *holds and let* $R = (R_1, R_2, \ldots, R_n)$ *where* $R_i = (c_{ii}a_i)^{-1}$. *Then there exists* $\epsilon > 0$ *such that*

$$\Sigma(s) = [s\hat{u}^*, s\hat{u}^* + (1-s)(\hat{R} + \epsilon\hat{1})]$$

is positively invariant for $0 \leq s \leq 1$ *and is a contracting rectangle for* $0 < s < 1$ *for* (7.1).

PROOF. Observe that $R_i > u_i^*$ so $su^* \leq u^* \leq su^* + (1-s)(R+\epsilon 1)$ for $0 \leq s \leq 1$ and $\epsilon > 0$. In order to show that $\Sigma(s)$ is a contracting rectangle for $0 < s < 1$, let $\phi \in \Sigma(s)$. If $\phi_i(0) = su_i^*$, then

$$f_i(\phi) \geq b_i s u_i^* \Big[1 - c_{ii}a_i s u_i^* - c_{ii}[su_i^* + (1-s)(R_i + \epsilon)](1-a_i)$$
$$- \sum_{j \neq i} c_{ij}[su_j^* + (1-s)(R_j + \epsilon)]\Big]$$
$$= b_i s u_i^* \Big[1 - s - c_{ii}(1-a_i)(1-s)(R_i + \epsilon) - \sum_{j \neq i} c_{ij}(1-s)(R_j + \epsilon)\Big]$$
$$= b_i s(1-s)u_i^* \Big[1 - c_{ii}(1-a_i)(R_i + \epsilon) - \sum_{j \neq i} c_{ij}(R_j + \epsilon)\Big]$$
$$= b_i s(1-s)u_i^* \Big[2 - a_i^{-1} - \sum_{j \neq i} c_{ij}(c_{jj}a_j)^{-1} + O(\epsilon)\Big] > 0,$$

where the inequality holds for sufficiently small $\epsilon > 0$. If $\phi_i(0) = su_i^* + (1-s)(R_i + \epsilon)$, then

$$f_i(\phi)$$
$$\leq b_i \phi_i(0)\Big[1 - c_{ii}a_i\Big(su_i^* + (1-s)((c_{ii}a_i)^{-1} + \epsilon)\Big) - c_{ii}(1-a_i)su_i^* - \sum_{j \neq i} c_{ij}su_j^*\Big]$$
$$= b_i \phi_i(0)[1 - s - c_{ii}a_i\Big(su_i^* + (1-s)((c_{ii}a_i)^{-1} + \epsilon)\Big) - c_{ii}(1-a_i)su_i^* + c_{ii}su_i^*]$$
$$= b_i \phi_i(0)[1 - s - c_{ii}a_i(1-s)((c_{ii}a_i)^{-1} + \epsilon)]$$
$$= -b_i(1-s)\phi_i(0)\epsilon c_{ii}a_i < 0.$$

This completes the proof. □

Lemma 7.4 leads immediately to the global stability of u^*.

THEOREM 7.5. *If* (7.4) *holds*, $\phi \in C_+$ *and* $0 \ll \phi(0)$, *then*
$$u(t,\phi) \to u^*, \quad t \to \infty.$$

PROOF. By Proposition 7.2 and $\phi(0) \gg 0$ it follows that $u(t,\phi) \gg 0$ for $t > 0$ and, since $z_i(t) \to (c_{ii}a_i)^{-1}$ as $t \to \infty$, there exists $T > 0$ and $\epsilon > 0$ such that $0 < u_i(t) < (c_{ii}a_i)^{-1} + \epsilon$ for $t > T$. Therefore, the result follows from Lemma 7.4 and Theorem 2.5. □

In the special case that $n = 2$, (7.1) generates a monotone dynamical system with respect to the ordering $\phi = (\phi_1, \phi_2) \leq_K \psi = (\psi_1, \psi_2)$ whenever $\phi_1 \leq \psi_1$ and $\phi_2 \geq \psi_2$.

8. Remarks and Discussion

The presentation of this chapter is largely based on Smith(1987a). However, the quasimonotone condition (Q) and results similar to Theorem 1.1 were discovered by a number of researchers at an earlier time. See Kunische and Schappacher(1979), Martin(1981) and Ohta(1981). The sufficient conditions for the positive invariance of C_+, given in Theorem 2.1, are a special case of more general invariance results of Seifert(1976). These conditions are easily seen to be necessary for the conclusion as well. See also Martin(1981).

The construction of quasimonotone comparison systems related to (0.1) when the latter has a positively invariant order interval follows the treatment in Martin and Smith(1991a) where it is carried out for reaction-diffusion systems with time delays, extending earlier work of Pozio(1982). These ideas seem to have first appeared in Conway and Smoller(1977) for reaction-diffusion systems and have been extended by Gardner(1980) and Brown(1980). See also Smoller(1983). A number of applications of the contracting rectangles argument can be found in the above-mentioned references. For extensions to convex sets other than rectangles, see Martin and Smith(1991a).

Theorem 3.4 which provides sufficient conditions for eventual strong monotonicity of the semiflow generated by a delay differential equation is due to Smith(1987a). Related results can be found in Martin(1981).

The principal results of section 5 are that the stability of an equilibrium of a cooperative and irreducible system of delay differential equations is determined by a real characteristic root (Theorem 5.1) and that this stability is the same as for an associated system of cooperative ordinary differential equations (Corollary 5.2). Actually, a more general result appears in Smith(1987a). The conditions (R) and (I) are not necessary for the conclusion that $s(L)$ is a root and that $\Re\lambda < s(L)$ for all other roots. These conditions are necessary for the other conclusions however. By assuming (I) and (R), we are able to give a simpler proof than Smith(1987a) which does not require special results about positive semigroups.

At the time of the writing of Smith(1987a), the author was unaware of work of Kerscher and Nagel(1984) which contains more general results on stability for abstract linear delay differential equations. Their results are based on a "Perron-Frobenius theory" for strongly continuous semigroups (Greiner(1981)) and the fact

8. REMARKS AND DISCUSSION

that the spectral bound for a positive semigroup of operators on a Banach lattice belongs to the spectrum of the generator (Batty and Robinson(1984)). This theory is very important for the theory of monotone dynamical systems since ultimately the stability of equilibria for nonlinear monotone systems is obtained by linearization. Linearization leads to a positive semigroup of operators. The problem then boils down to showing that the growth bound of a positive semigroup agrees with the spectral bound of its infinitesimal generator. The reader is referred to the paper of Kerscher and Nagel(1984) and the references therein for many interesting results in this direction. Theorem 5.1 is a special case of their results.

There is a large literature on the biochemical control system (6.1) of section 6. For more general results, different models and many references to the literature as of 1987, the reader may consult Smith(1987b). See also Smith and Thieme(1991b) for more general results.

Section 7 follows Martin and Smith(1991a). The results of this section can be extended to reaction-diffusion systems with delays and to systems that are not necessarily competitive. See Martin and Smith(1991b). As noted at the end of section 7, equation (7.1) generates a monotone semiflow when $n = 2$. This fact is exploited in Martin and Smith(1991a) to give sufficient conditions for one of the populations to eliminate its rival.

The delay differential equations considered in this chapter have bounded time delays. However, unbounded and even infinite delays arise naturally in the applications in the form of Volterra integrodifferential equations. Recently, Wu(1992) has extended the theory developed here to these kind of systems. Equations of neutral type are treated in Wu and Freedman(1991).

Many of the results of this chapter can be extended to reaction-diffusion systems with time delays. See Martin and Smith(1990,1991a,b).

CHAPTER 6

Nonquasimonotone Delay Differential Equations

0. Introduction

In the previous chapter we saw that the scalar delay differential equation given by
$$x'(t) = g(x(t), x(t-r)) \tag{0.1}$$
generates an eventually strongly monotone semiflow on the space C of continuous functions on $[-r, 0]$ with the usual pointwise ordering provided
$$\frac{\partial g}{\partial y}(x, y) > 0 \tag{0.2}$$
holds. In other words, the right hand side of (0.1) must be strictly increasing in the delayed argument. This is a severe restriction on (0.1) and it is natural to ask if it can be relaxed. As the quasimonotone condition (Q) of the previous chapter is necessary for the semiflow to be monotone in the sense of the usual pointwise ordering, there is no hope in relaxing (0.2) without identifying a new cone or partial ordering. The aim of the present chapter is to do just that. A new more restrictive cone is identified which allows the theory of Chapters 1 and 2 to be applied under less restrictive conditions than (0.2). In particular, we will see that monotonicity of g in the undelayed argument x can mitigate the failure of (0.2) to hold. But even if this too fails, not all is lost. If the delay is small enough compared to the Lipschitz constant for (0.1), the theory developed in this chapter applies. Small delays are harmless, so goes the folklore of the subject, and we prove this rigorously.

A novelty of the theory constructed in this chapter is that the full power of the strong order preserving hypothesis is required since it turns out that the semiflow is not strongly monotone nor eventually strongly monotone.

The presentation follows a familiar pattern. In the next section, the new ordering is introduced and sufficient conditions are given for a delay differential equation to preserve this ordering. Sufficient conditions for the strong order preserving property to hold are developed in section 2 and this leads to the main result, Theorem 3.1. As usual, the conclusion is that the generic orbit converges to equilibrium. The Krein-Rutman theorem plays its customary role in stability considerations as described in section 4. In section 5, the theory is applied to a model of a fly population. Here, it is shown that the methods developed in this chapter allow a significant improvement over those of the previous chapter in the main result.

1. The Exponential Ordering

We consider the more general equation
$$x'(t) = f(x_t) \tag{1.1}$$

using the notation of the previous chapter. For simplicity, we restrict attention to the case that (1.1) is a scalar equation in this chapter. Some remarks on the extension to systems of delay equations are made in the remarks and discussion section at the end of the chapter. We assume that $f: D \to \mathbb{R}$ is continuous and satisfies a Lipschitz condition on each compact subset of D where D is an open subset of C.

Let us introduce the new cones. Consider the family of sets parametrized by $\mu \geq 0$ given by

$$\tilde{K}_\mu = \{\phi \in C : \phi \geq 0 \text{ and } \phi(s)e^{\mu s} \text{ is nondecreasing on } [-r, 0]\}.$$

It is easy to see that $\tilde{K}_\mu$ is a closed cone in C. It generates a partial order on C, which we write as $\leq_\mu$, defined as

$$\phi \leq_\mu \psi \iff \phi \leq \psi \text{ and } (\psi(s) - \phi(s))e^{\mu s} \text{ is nondecreasing on } [-r, 0].$$

This ordering is sometimes called the *exponential ordering*. We write $\phi <_\mu \psi$ whenever $\phi \leq_\mu \psi$ and $\phi \neq \psi$. Observe that if $\phi <_\mu \psi$, then the monotonicity of $e^{\mu s}(\psi(s) - \phi(s))$ implies that either $\phi(s) < \psi(s)$ for $s \in [-r, 0]$ or their exists $s_0 \in [-r, 0)$ such that $\phi(s) = \psi(s)$ for $s \in [0, s_0]$ and $\phi(s) < \psi(s)$ for $s \in (s_0, 0]$.

If ϕ and ψ are differentiable functions then $\phi \leq_\mu \psi$ if and only if

$$\phi \leq \psi \quad \text{and} \quad (\psi - \phi)' + \mu(\psi - \phi) \geq 0.$$

This last differential inequality is the key to understanding why the exponential ordering is useful. Note that if x and y are real numbers such that $x \leq y$, then the constant functions $\hat{x}$ and $\hat{y}$ satisfy $\hat{x} \leq_\mu \hat{y}$. Thus the constant functions retain the same ordering as before. In particular, the set of equilibria of (1.1) is totally ordered by $\leq_\mu$.

It is apparent that $\tilde{K}_\mu \subset C_+$ and this implies that $\tilde{K}_\mu$ is normal (see beginning of Chapter 2). For if $0 \leq_\mu \phi \leq_\mu \psi$ then $0 \leq \phi \leq \psi$ so $|\phi| \leq |\psi|$ because C_+ is normal. In fact, $\tilde{K}_\mu$ is quite narrow. It has empty interior in C since every neighborhood in C of a nondecreasing function contains a function lacking this property. However the restrictive nature of the cone is an asset when it comes to determining sufficient conditions for the semiflow to preserve it since there is more information in the inequality $\leq_\mu$ to work with. We now turn to this problem.

If $x_t(\phi) \leq_\mu x_t(\psi)$ holds for $t \geq 0$ then it must be the case that $(x'(t, \psi) - x'(t, \phi)) + \mu(x(t, \psi) - x(t, \phi)) \geq 0$. Putting $t = 0$ and using (1.1), we see that the following condition is necessary for (1.1) to generate a monotone semiflow with respect to $\leq_\mu$.

(M_μ) Whenever $\phi, \psi \in D$ satisfy $\phi \leq_\mu \psi$, then $\mu(\psi(0) - \phi(0)) + f(\psi) - f(\phi) \geq 0$.

The condition (M_μ) is also sufficient for (1.1) to generate a monotone semiflow with respect to the ordering $\leq_\mu$ as we now show.

THEOREM 1.1. *Let (M_μ) hold and $\phi \leq_\mu \psi$ $(\phi <_\mu \psi)$. Then*

$$x_t(\phi) \leq_\mu x_t(\psi) \quad (x_t(\phi) <_\mu x_t(\psi))$$

for all $t \geq 0$ such that both solutions are defined.

PROOF. Fix $\epsilon > 0$ and let $f_\epsilon : D \to \mathbb{R}$ be defined by $f_\epsilon(\phi) = f(\phi) + \epsilon\phi(0)$. Let $x(t, \phi, \epsilon)$ denote the maximally defined solution of
$$x'(t) = f_\epsilon(x_t), \quad t \geq 0, \quad x(\theta) = \phi(\theta), \quad -r \leq \theta \leq 0.$$
As usual, we will first establish the result for the solutions $x(t, \phi, \epsilon)$ and $x(t, \psi, \epsilon)$ and then let $\epsilon \to 0$. Let $\phi <_\mu \psi$ so $\phi(0) < \psi(0)$. Then by (M_μ) we have
$$\frac{d^+}{dt}|_{t=0} e^{\mu t}[x(t, \psi, \epsilon) - x(t, \phi, \epsilon)]$$
$$\geq (\mu + \epsilon)[\psi(0) - \phi(0)] + f(\psi) - f(\phi)$$
$$\geq \epsilon[\psi(0) - \phi(0)] > 0.$$
It follows that $x(t, \phi, \epsilon) \leq x(t, \psi, \epsilon)$ and $e^{\mu t}(x(t, \psi, \epsilon) - x(t, \phi, \epsilon))$ is nondecreasing on an interval $[0, \delta)$ for some $\delta > 0$. Let $t_1 > 0$ be the largest number such that $x(t, \phi, \epsilon) \leq x(t, \psi, \epsilon)$ and $e^{\mu t}(x(t, \psi, \epsilon) - x(t, \phi, \epsilon))$ is monotone nondecreasing for $t \in [0, t_1)$. If t_1 is strictly less than the right endpoint of the intersection of the maximal intervals of existence of the two solutions, then continuity implies $x(t_1, \phi, \epsilon) \leq x(t_1, \psi, \epsilon)$. Furthermore, equality cannot hold since then monotonicity implies that $x(t, \phi, \epsilon) = x(t, \psi, \epsilon)$ for all t, in contradiction to the maximality of t_1. Thus, $x(t_1, \psi, \epsilon) > x(t_1, \phi, \epsilon)$. By (M_μ) and the fact that $x_{t_1}(\phi, \epsilon) \leq_\mu x_{t_1}(\psi, \epsilon)$, we have
$$\frac{d^+}{dt}|_{t=t_1} e^{\mu t}[x(t, \psi, \epsilon) - x(t, \phi, \epsilon)]$$
$$\geq e^{\mu t_1}\{(\mu + \epsilon)[x(t_1, \psi, \epsilon) - x(t_1, \phi, \epsilon)] + f(x_{t_1}(\psi, \epsilon)) - f(x_{t_1}(\phi, \epsilon))\}$$
$$\geq \epsilon e^{\mu t_1}[x(t_1, \psi, \epsilon) - x(t_1, \phi, \epsilon)] > 0.$$
Hence $x(t, \phi, \epsilon) \leq x(t, \psi, \epsilon)$ and $e^{\mu t}(x(t, \psi, \epsilon) - x(t, \phi, \epsilon))$ is nondecreasing beyond $t = t_1$, in contradiction to the maximality of t_1. Therefore, we conclude that these properties hold so long as both solutions are defined.

Now suppose that $x(t, \psi, 0)$ and $x(t, \phi, 0)$ are both defined on $0 \leq t \leq s$. Since $f_\epsilon \to f$ uniformly on bounded subsets of D, it follows from Hale(1977), Theorem 2.2.2, that $x(t, \psi, \epsilon)$ and $x(t, \phi, \epsilon)$ are defined on $0 \leq t \leq s$ for all sufficiently small $\epsilon > 0$ and $x(t, \psi, \epsilon) \to x(t, \psi, 0)$ and $x(t, \phi, \epsilon) \to x(t, \phi, 0)$ uniformly on $0 \leq t \leq s$. Taking limits as $\epsilon \to 0$ in the inequality $x(t, \phi, \epsilon) \leq x(t, \psi, \epsilon)$ and the inequality $e^{\mu t_1}(x(t_1, \psi, \epsilon) - x(t_1, \phi, \epsilon)) \leq e^{\mu t_2}(x(t_2, \psi, \epsilon) - x(t_2, \phi, \epsilon))$, where $0 \leq t \leq s$ and $0 \leq t_1 \leq t_2 \leq s$ establishes that $x_t(\phi) \leq_\mu x_t(\psi)$ for $t \geq 0$ such that both solutions are defined.

If $\phi <_\mu \psi$ then it follows that $\phi(0) < \psi(0)$ by the monotonicity of $e^{\mu \theta}(\psi(\theta) - \phi(\theta))$ for $-r \leq \theta \leq 0$. From the arguments above $e^{\mu t}(x(t, \psi) - x(t, \phi))$ is nondecreasing for $t \geq 0$ such that both solutions are defined and since this quantity is positive at $t = 0$ it must remain positive. Thus, $x(t, \phi) < x(t, \psi)$ for $t \geq 0$ such that both solutions are defined. This establishes that $x_t(\phi) <_\mu x_t(\psi)$. □

Sufficient conditions for (M_μ) to hold for some $\mu \geq 0$ are given below. In most cases, the particular value of $\mu \geq 0$ such that (M_μ) holds is not of interest. Perhaps one of the more remarkable of these conditions is given immediately below.

(L^-) There exists $L > 0$ such that $f(\psi) - f(\phi) \geq -L|\psi - \phi|$ whenever $\phi, \psi \in D$ and $\phi \leq \psi$.

PROPOSITION 1.2. *If f satisfies (L^-) and if*
$$erL \leq 1 \qquad (1.2)$$
then (M_μ) holds for some $\mu \geq 0$. In particular, if f is Lipschitz on D with constant $L > 0$ and if (1.2) holds, then (M_μ) holds for some $\mu \geq 0$.

PROOF. If $\phi \leq_\mu \psi$ then $e^{\mu s}(\psi(s) - \phi(s)) \leq \psi(0) - \phi(0)$ so $|\psi - \phi| \leq e^{\mu r}(\psi(0) - \phi(0))$. Consequently, if (L^-) holds then
$$\mu(\psi(0) - \phi(0)) + f(\psi) - f(\phi) \geq \mu(\psi(0) - \phi(0)) - L|\psi - \phi|$$
$$\geq (\mu - Le^{\mu r})(\psi(0) - \phi(0))$$
$$\geq 0$$
provided $\mu - Le^{\mu r} \geq 0$. It is now an exercise in calculus to show that the latter holds for some $\mu \geq 0$ if and only if (1.2) holds. □

If the delay r is small compared to the Lipschitz constant for (1.1), in the sense that (1.2) holds, then (1.1) generates a monotone semiflow with respect to an ordering $\leq_\mu$. The folklore in the theory of delay equations says that small delays are harmless and can be ignored. The theory developed in this chapter will show rigorously that this is the case. It will even provide an estimate of how small the delay must be to be harmless.

Let us see how the new condition (M_μ) applies to (0.1). Suppose that g satisfies

(G) There exists L_1 and L_2 such that whenever $x_1 \leq x_2$ and $y_1 \leq y_2$, then
$$g(x_2, y_2) - g(x_1, y_1) \geq L_1(x_2 - x_1) + L_2(y_2 - y_1).$$

The next result gives conditions on L_1 and L_2 such that (M_μ) holds for some $\mu \geq 0$.

PROPOSITION 1.3. *(M_μ) holds for (0.1) for some $\mu \geq 0$ if (G) holds and one of the following hold:*
(a) $L_2 \geq 0$, or
(b) $L_2 < 0$ but $L_1 + L_2 \geq 0$, or
(c) $L_2 < 0$, $L_1 + L_2 < 0$, $r|L_2| < 1$, and $rL_1 - \log(r|L_2|) \geq 1$.

PROOF. If $\phi \leq_\mu \psi$, then $\psi(-r) - \phi(-r) \leq e^{\mu r}(\psi(0) - \phi(0))$. Therefore,
$$\mu(\psi(0) - \phi(0)) + g(\psi(0), \psi(-r)) - g(\phi(0), \phi(-r)) \geq$$
$$(\mu + L_1)(\psi(0) - \phi(0)) + L_2(\psi(-r) - \phi(-r)) \geq$$
$$(\mu + L_1)(\psi(0) - \phi(0)) + L_2^- e^{\mu r}(\psi(0) - \phi(0)) \geq$$
$$(\mu + L_1 + L_2^- e^{\mu r})(\psi(0) - \phi(0)) \geq 0$$
provided $\mu + L_1 + L_2^- e^{\mu r} \geq 0$. Here, $L_2^- = \min\{0, L_2\}$. The latter holds if one of (a), (b) or (c) holds. □

We may view L_1 as the infimum of $\frac{\partial g}{\partial x}(x, y)$ and L_2 as the infimum of $\frac{\partial g}{\partial y}(x, y)$ where the infimum is taken over all (x, y). According to (a), (0.1) generates a monotone semiflow when g is nondecreasing with respect to the delayed argument y, just as in the previous chapter, except that the cone is different. However, if g

fails to be nondecreasing with respect to y but at least $L_1 + L_2 \geq 0$ then (0.1) still generates a monotone semiflow by (b). In other words, monotonicity of g in the undelayed variable can offset the failure of g to be monotone in the delayed variable. If $L_2 < 0$ and $L_1 + L_2 < 0$, not all is lost. According to (c), (0.1) still generates a monotone semiflow if the delay is small enough. The delay r in (0.1) can be made to be 1 by a change in time scale which results in r multiplying g. Therefore, it is not surprising that the product rL_i appears in (c). It seems apparent that the new ordering requires less restrictive conditions to be placed on g.

Finally, consider the general autonomous linear delay equation

$$x'(t) = Lx_t \tag{1.3}$$

where L is a bounded linear functional on C. As such, it has a representation in terms of a signed Borel measure ν as

$$L\phi = \int_{[-r,0]} \phi(s) d\nu(s).$$

We seek necessary and sufficient conditions on ν for (1.3) to generate a monotone semigroup with respect to the ordering $\leq_\mu$.

PROPOSITION 1.4. (M_μ) holds for (1.3) if and only if

$$\mu + \int_H e^{-\mu s} d\nu(s) \geq 0 \tag{1.4}$$

for $H = [-r, 0]$ and for $H = (\theta, 0]$ for each θ satisfying $-r \leq \theta < 0$.

PROOF. Since L is linear, (M_μ) is equivalent to showing that $L\phi + \mu\phi(0) \geq 0$ whenever $0 \leq_\mu \phi$. If $K\phi \equiv L\phi + \mu\phi(0)$ then it suffices to show that $K\phi \geq 0$ whenever $0 \leq_\mu \phi$. We first show the necessity of (1.4). Since $\phi(s) = e^{-\mu s} \in \tilde{K}_\mu$ it follows that $K\phi \geq 0$. This is just (1.4) with $H = [-r, 0]$. Now let χ be the characteristic function of $H = (\theta, 0]$ and ϕ_n be a sequence of continuous nondecreasing functions which vanish on $[-r, \theta]$, $\phi_n(0) = 1$ and $\phi_n \to \chi$ pointwise on $[-r, 0]$ as $n \to \infty$. Then $\psi_n \equiv \phi_n e^{-\mu s} \in \tilde{K}_\mu$ so $K\psi_n \geq 0$. Therefore,

$$0 \leq \int_{[-r,0]} \phi_n(s) e^{-\mu s} d\nu(s) + \mu.$$

Letting $n \to \infty$ we get (1.4) with $H = (\theta, 0]$.

Now suppose that (1.4) holds for the indicated sets H and fix $\phi \in \tilde{K}_\mu$. We must show that $K\phi \geq 0$. K is represented by the measure $\eta = \nu + \mu\delta_0$, where δ_0 is the Dirac measure with mass at 0. Let ξ be the measure defined by $\xi(H) = \int_H e^{-\mu s} d\eta(s)$. If $g(s) = \phi(s) e^{\mu s}$ then g is nonnegative and nondecreasing and

$$K\phi = \int_{[-r,0]} \phi(s) e^{\mu s} e^{-\mu s} d\eta(s) = \int_{[-r,0]} g(s) d\xi(s).$$

This may be rewritten as

$$K\phi = g(0)\xi([-r,0]) - \int_{[-r,0]} [g(0) - g(s)]d\xi(s)$$

$$= \phi(0)[\mu + \int_{[-r,0]} e^{-\mu s} d\nu(s)] - \int_{[-r,0]} \left(\int_{[s,0)} dg(t)\right) d\xi(s)$$

$$= \phi(0)[\mu + \int_{[-r,0]} e^{-\mu s} d\nu(s)] - \int_{[-r,0)} \left(\int_{[-r,t]} d\xi(s)\right) dg(t)$$

$$= \phi(0)[\mu + \int_{[-r,0]} e^{-\mu s} d\nu(s)] - \int_{[-r,0)} \left(\int_{[-r,t]} e^{-\mu s} d\nu(s)\right) dg(t)$$

$$= \phi(0)[\mu + \int_{[-r,0]} e^{-\mu s} d\nu(s)]$$
$$- \int_{[-r,0)} \left(\int_{[-r,0]} e^{-\mu s} d\nu(s) - \int_{(t,0]} e^{-\mu s} d\nu(s)\right) dg(t)$$

$$= \phi(0)\mu + e^{-\mu r}\phi(-r) \int_{[-r,0]} e^{-\mu s} d\nu(s) + \int_{[-r,0)} \left(\int_{(t,0]} e^{-\mu s} d\nu(s)\right) dg(t).$$

Now using (1.4) with $H = [-r, 0]$ and $H = (t, 0]$ we have

$$K\phi \geq \phi(0)\mu - \mu\phi(-r)e^{-\mu r} - \int_{[-r,0)} (\mu) dg(t)$$
$$= \phi(0)\mu - \mu\phi(-r)e^{-\mu r} - \mu(\phi(0) - e^{-\mu r}\phi(-r))$$
$$= 0.$$

This completes the proof. □

As a simple example, consider the equation

$$x'(t) = ax(t) + bx(t-r).$$

The linear functional $L\phi = a\phi(0) + b\phi(-r)$ is represented by the measure $\nu = a\delta_0 + b\delta_{-r}$. The condition (1.4) holds if and only if

$$0 \leq \mu + a + b^- e^{\mu r}$$

where $b^- = \min\{b, 0\}$. This is just the condition encountered in the proof of Proposition 1.3. It is satisfied for some $\mu \geq 0$ if and only if one of (a),(b) or (c) of Proposition 1.3 hold where a replaces L_1 and b replaces L_2.

REMARK 1.1. If f is continuously differentiable and for each $\phi \in D$, the derivative $df(\phi)$ is represented by the Borel measure $\nu(\phi)$, then (M_μ) holds for f provided $\nu(\phi)$ satisfies the hypotheses of Proposition 1.4 and D is order convex. The proof is the same as for Lemma 5.3.3.

2. The Strong Order Preserving Property

The strong order preserving property must be established in order to apply the theorems of the first two chapters. This is the aim of the present section and it will not be as easy as in previous chapters because, as noted in the previous section, the cone $\tilde{K}_\mu$ has empty interior in C. We cannot, therefore, expect to show that the semiflow Φ generated by (1.1) is eventually strongly monotone on C. However, it was pointed out in Remark 1.1 of Chapter 1 that if a Banach subspace Z of C can be found such that the inclusion map $i : Z \to C$ is continuous and such that $K_\mu = Z \cap \tilde{K}_\mu$ has nonempty interior in Z, then the strong order preserving property (SOP) holds if two things can be shown. These are: (1) there exists $t_0 > 0$ such that Φ_{t_0} maps C continuously into Z and (2) whenever $\phi \leq_\mu \psi$ holds in C then $\Phi_{t_0}(\phi) \ll_\mu \Phi_{t_0}(\psi)$ holds in Z, where $\ll_\mu$ is the strong ordering on Z generated by $\text{Int} K_\mu$. This program will be carried out with $Z = C_L$, the Banach space of Lipschitz functions on $[-r, 0]$ with the norm

$$|\phi|_{Lip} = |\phi| + \text{Lip}(\phi)$$

where

$$\text{Lip}(\phi) = \sup\left\{\left|\frac{\phi(s) - \phi(t)}{s - t}\right| : s \neq t, \quad s, t \in [-r, 0]\right\}.$$

It is clear that the inclusion map $i : C_L \to C$ is continuous. If $\phi \in C_L$ then it is absolutely continuous, $\phi'(s)$ exists for almost all (a.e.) $s \in [-r, 0]$ with respect to Lebesgue measure and

$$|\phi'(s)| \leq \text{Lip}(\phi)$$

for all s for which $\phi'(s)$ exists.

As a consequence of the relative smoothness of functions in C_L, elements of the cone K_μ in C_L can be characterized in terms of their derivative.

$$K_\mu = \tilde{K}_\mu \cap C_L = \{\phi \in C_L : \phi \geq 0 \text{ and } \phi' + \mu\phi \geq 0 \text{ a.e. in } [-r, 0]\}.$$

The key question is whether K_μ has nonempty interior in C_L. It does and in order to identify those functions belonging to $\text{Int} K_\mu$ we recall that if ψ is defined and measurable on $\text{dom}(\phi) \subset [-r, 0]$, then the *essential infimum* of ψ is defined by

$$\text{ess inf}\,\psi \equiv \sup\{y : m(\{s \in \text{dom}(\phi) : \psi(s) < y\}) = 0\}$$

where m denotes Lebesgue measure. If $e = \text{ess inf}\,\psi$ then $\psi(s) \geq e$ almost everywhere on $\text{dom}(\phi)$. In the case of interest here, $\psi = \phi' + \mu\phi$ for some $\phi \in C_L$ and therefore the complement of $\text{dom}(\psi)$ in $[-r, 0]$ has zero Lebesgue measure and ψ is bounded. Now we can characterize the interior of K_μ in C_L.

LEMMA 2.1. *K_μ has nonempty interior in C_L given by*

$$\text{Int} K_\mu = \{\phi \in C_L : \phi \gg 0 \text{ and ess inf}(\phi' + \mu\phi) > 0\}.$$

PROOF. Suppose that $\phi \in C_L$ satisfies $\inf_{[-r,0]} \phi(s) = q > 0$ and $p = \text{ess inf}(\phi' + \mu\phi) > 0$. Let $\epsilon > 0$ satisfy $\epsilon < p/(1+\mu)$ and $\epsilon < q$. If $\psi \in C_L$ satisfies $|\psi - \phi|_{Lip} < \epsilon$, then $-\epsilon < \psi'(s) - \phi'(s) < \epsilon$ a.e. and $-\epsilon < \psi(s) - \phi(s) < \epsilon$ on $[-r, 0]$. Hence, $\psi(s) > \phi(s) - \epsilon \geq q - \epsilon \geq 0$ and $\psi'(s) + \mu\psi(s) \geq \phi'(s) - \epsilon + \mu(\phi(s) - \epsilon) \geq p - (1+\mu)\epsilon > 0$ holds a.e. on $[-r, 0]$. Therefore, $\psi > 0$ and $\text{ess inf}(\psi' + \mu\psi) > 0$ and consequently $\psi \in K_\mu$. We conclude that $\phi \in \text{Int} K_\mu$.

Conversely, if $\phi \in \text{Int} K_\mu$ then for all small $\epsilon > 0$, $\phi - \epsilon\psi \in K_\mu$ where $\psi(s) = 1 + (1/2r)s$. Therefore, $\phi(s) - \epsilon\psi(s) \geq 0$ and consequently $\phi \gg 0$. Also, $\phi'(s) - (\epsilon/2r) + \mu(\phi(s) - \epsilon\psi(s)) \geq 0$ a.e. so $\phi'(s) + \mu\phi(s) \geq (\epsilon/2r) + \epsilon\mu\psi(s) \geq \epsilon(1/2r + \mu/2)$ a.e.. It follows that ess $\inf(\phi' + \mu\phi) \geq \epsilon(1/2r + \mu/2)$. Therefore, ϕ has the properties described in the Lemma. □

Because the cone K_μ has nonempty interior in C_L, the strong ordering can be defined. We write
$$\phi \ll_\mu \psi \iff \psi - \phi \in \text{Int} K_\mu.$$

Now that we have identified the Banach subspace C_L of C and shown that the cone K_μ has nonempty interior in C_L, we need to establish that the two conditions (1) and (2), mentioned at the beginning of the section, hold in this setting. Returning to the equation (1.1), it will be assumed throughout the remainder of the chapter that the solution corresponding to the initial data $x_0 = \phi \in D$ is defined for all $t \geq 0$. In this case, (1.1) generates a semiflow on C as usual by
$$\Phi_t(\phi) = x_t(\phi).$$

The first condition (1), the continuity of Φ_{t_0} as a map from C to C_L for $t_0 \geq r$, is established next.

LEMMA 2.2. *If $t_0 \geq r$ then the map $\phi \to x_{t_0}(\phi)$ is continuous as a map from C into C_L.*

PROOF. Standard theory of delay equations implies that if $\phi_n \to \phi$ in C, then $x(t, \phi_n) \to x(t, \phi)$ as $n \to \infty$, uniformly on $-r \leq t \leq t_0$. See Hale(1977). Since $t_0 \geq r$ and f is continuous, it follows that $x'_{t_0}(\phi_n)(s) = x'(t_0 + s, \phi_n) = f(x_{t_0+s}(\phi_n)) \to f(x_{t_0+s}(\phi)) = x'_{t_0}(\phi)(s)$ as $n \to \infty$, uniformly for $-r \leq s \leq 0$. Thus, $x'_{t_0}(\phi_n) \to x'_{t_0}(\phi)$ in C as $n \to \infty$. Since Lip$(\phi) \leq |\phi'|$ if ϕ is continuously differentiable, it follows that $x_{t_0}(\phi_n) \to x_{t_0}(\phi)$ in C_L. □

In order to establish the second condition (2) for Φ to satisfy SOP, a slightly stronger condition than (M_μ) is required. Basically, the inequality in (M_μ) must be strict in case $\phi <_\mu \psi$.

(SM_μ) there exists $\mu \geq 0$ such that whenever $\phi, \psi \in C$ satisfy $\phi <_\mu \psi$, then
$$\mu(\psi(0) - \phi(0)) + f(\psi) - f(\phi) > 0.$$

The main result of this section provides sufficient conditions for Φ to satisfy SOP.

THEOREM 2.3. *If (SM_μ) holds and $\phi, \psi \in C$ satisfy $\phi <_\mu \psi$, then $x_t(\phi) \ll_\mu x_t(\psi)$ holds in C_L for all $t \geq r$. In particular, the semiflow Φ is SOP on C.*

PROOF. Since (SM_μ) implies that (M_μ) holds, we conclude from Theorem 1.1 that $x_t(\phi) <_\mu x_t(\psi)$ for all $t \geq 0$ and consequently $x(t, \phi) < x(t, \psi)$ holds for all $t \geq 0$. For $t \geq 0$ we have, by (SM_μ), that
$$x'(t,\psi) - x'(t,\phi) + \mu(x(t,\psi) - x(t,\phi)) = \mu[x(t,\psi) - x(t,\phi)] + f(x_t(\psi)) - f(x_t(\phi)) > 0.$$
The first conclusion follows immediately from the strict inequalities and continuity. The final assertion follows from Remark 1.1 of Chapter 1, Lemma 2.1 and Lemma 2.2. □

The next several remarks provide some sufficient conditions for (SM_μ) to hold. They are all rather minor modifications of the sufficient conditions for (M_μ) to hold which were noted following Theorem 1.1.

REMARK 2.1. It is a routine exercise to check that if f satisfies (L^-) on D and

$$erL < 1 \tag{2.1}$$

then (1.1) satisfies (SM_μ) for some $\mu \geq 0$. Only a slight modification of the proof of Proposition 1.2 is required (note that $\phi(0) < \psi(0)$ holds if $\phi <_\mu \psi$). Inequality (2.1) gives a quantitative estimate of how small a delay must be to be harmless.

REMARK 2.2. Similarly, if g satisfies (G), then (SM_μ) holds for some $\mu \geq 0$ for (0.1) provided one of the following holds:
(i) $L_2 \geq 0$, or
(ii) $L_2 < 0$ but $L_1 + L_2 > 0$, or
(iii) $L_2 < 0$, $L_1 + L_2 = 0$, and $r|L_2| < 1$, or
(iv) $L_2 < 0$, $L_1 + L_2 < 0$, $r|L_2| < 1$, and $rL_1 - \log(r|L_2|) > 1$.

It is perhaps worth reemphasizing the remarks made in connection with Proposition 1.3 which gave analogous conditions for (M_μ) to hold for (0.1). According to Remark 2.2, (SM_μ) holds for some $\mu \geq 0$ if g is nondecreasing in the delayed variable ((i) above) but if that fails, monotonicity of g in the undelayed variable x may permit salvaging the SOP property ((ii) above). If both conditions fail, but the delay is suitably small, then the SOP property still holds.

(SM_μ) holds for the autonomous linear system (1.3) if the inequality (1.4) is strict for the indicated sets H.

3. Generic Convergence to Equilibrium

We are now in position to apply the results of Chapters 1 and 2 to the scalar delay differential equation

$$x'(t) = f(x_t) \tag{3.1}$$

where $f : D \to \mathbb{R}$ is continuous and satisfies a Lipschitz condition on each compact subset of D. It will be assumed that for each $\phi \in D$ the solution of (3.1) satisfying $x_0 = \phi$ is defined for all $t \geq 0$. The same compactness assumption (T) used in the previous chapter will also be required. It is restated below for convenience.

(T) f maps bounded subsets of D to bounded subsets of $\mathbb{R}$. For each $\phi \in D$, $x(t, \phi)$ is bounded for $t \geq 0$. Moreover, for each compact subset A of D, there exists a closed and bounded subset $B = B(A)$ of D such that for each $\phi \in A$, $x_t(\phi) \in B$ for all large t.

The main result of this chapter follows. It says that the generic solution converges to equilibrium.

THEOREM 3.1. *Suppose that (SM_μ) and (T) hold for (3.1). Then the set of convergent points in D contains an open and dense subset. If E consists of a single point, then it attracts all solutions of (3.1). If D is order convex and E consists of two points, then all solutions converge to one of these.*

PROOF. We apply Theorem 1.4.3 and Remark 4.2 of Chapter 1. Since D is open in C, if $\phi \in D$ then $\phi - \frac{1}{n}\hat{1}$ approximates ϕ from below. Here, we use that $\hat{0} <_\mu \hat{1}$. Therefore, every point of D can be approximated from below. As shown in the proof of Theorem 5.4.1, (T) implies the compactness assumption (C) of Chapter 1 holds. By Theorem 2.3, the semiflow Φ defined by $\Phi_t(\phi) = x_t(\phi)$ for $t \geq 0$ and $\phi \in D$ is SOP in the space C. Therefore, Theorem 1.4.3 implies that the set of quasiconvergent points contains an open and dense subset of D. It has already been remarked that the set E of equilibria is totally ordered by $\leq_\mu$. Consequently, by Remark 4.2 of Chapter 1 we may conclude that every quasiconvergent point is a convergent point. Therefore, the set of convergent points contains an open and dense subset. The last two assertions follow in the familiar way from Theorem 2.3.1 and Theorem 2.3.2. □

In contrast to Theorem 5.4.2, very minimal smoothness assumptions were required of f in Theorem 3.1. This was possible because we restricted consideration to scalar equations in this chapter and could take advantage of the fact that the set E is totally ordered by $\leq_\mu$. This fact allowed us to use Theorem 1.4.3 rather than Theorem 2.4.7, which requires more smoothness.

4. Stability of Equilibria

Suppose f satisfies (M_μ) and is continuously differentiable on D, and let v be an equilibrium of (3.1). The reader may recall that in the previous chapter we showed that the stability properties of v as an equilibrium of (3.1) are mimicked by the stability properties of v as an equilibrium of the ordinary differential equation obtained from (3.1) by ignoring the delay. A similar result will be obtained here.

Assume that $f(\hat{v}) = 0$. The linear variational equation corresponding to v is

$$y'(t) = Ly_t, \quad L = df(\hat{v}) \tag{4.1}$$

If f satisfies (M_μ) then it is easily shown that L satisfies $L\phi + \mu\phi(0) \geq 0$ whenever $0 \leq_\mu \phi$ holds. However, we will assume the strong inequality holds:

$$L\phi + \mu\phi(0) > 0 \text{ whenever } 0 <_\mu \phi. \tag{4.2}$$

Necessary and sufficient conditions for (4.2) to hold in terms of the Borel measure ν representing L were noted at the end of section 2. The characteristic equation corresponding to (4.1) (see Chapter 5, section 5) is given by

$$\lambda = L(e_\lambda), \quad e_\lambda(s) = e^{\lambda s}. \tag{4.3}$$

A complex number λ is a root of (4.3) if and only if (4.1) has an exponential solution $y(t) = e^{\lambda t}$. The stability modulus of L is

$$s(L) = \sup\{\Re\lambda : \lambda \text{ is a root of } (4.3)\}.$$

As noted in the previous chapter, v is asymptotically stable if $s(L) < 0$ and unstable if $s(L) > 0$. We now show that if L satisfies (4.2) then $s(L)$ is a root of (4.3) and all other roots have strictly smaller real part. In other words, the stability of v is determined by a real root of (4.3).

THEOREM 4.1. *Let L satisfy* (4.2). *Then $s(L)$ is a root of* (4.3) *and every other root has strictly smaller real part. Furthermore, $s(L) > -\mu$.*

PROOF. Let $\{T(t)\}_{t\geq 0}$ be the strongly continuous semigroup of bounded linear operators on C defined by $T(t)\phi = y_t(\phi)$ where $y(t,\phi)$ is the solution of (4.1) satisfying $y_0 = \phi$. $T(r)$ is a compact operator and its point spectrum consists of $e^{\lambda r}$ where λ is a root of (4.3). As $T(r)$ maps C into C_L, the corresponding eigenvectors belong to C_L. The space C_L is positively invariant under (4.1) but the restriction $T_L(t)$ of $T(t)$ to C_L is not a strongly continuous semigroup since the Lipschitz constant of $T_L(t)\phi$ can vary discontinuously at $t = 0$ for fixed $\phi \in C_L$. However, $T_L(r)$ is a compact operator on C_L. It is also strongly positive with respect to the cone K_μ since (4.2) and Theorem 2.3 imply that if $0 <_\mu \phi$ holds in C_L then $0 \ll_\mu y_r(\phi)$. By the Krein-Rutman Theorem (Theorem 2.4.1), the spectral radius ρ of $T_L(r)$ is a positive eigenvalue of algebraic multiplicity one and there is a corresponding eigenvector ψ satisfying $0 \ll_\mu \psi$. Furthermore, all other eigenvalues of $T_L(r)$ have strictly smaller modulus. But the set of eigenvalues of $T(r)$ and of $T_L(r)$ are the same, as are the generalized eigenspaces corresponding to each eigenvalue, so ρ is the spectral radius of $T(r)$ and all other eigenvalues are strictly smaller in modulus. Hence, $\rho = e^{\lambda r}$ for some root λ of (4.3). Since ρ is an eigenvalue of algebraic multiplicity one, it must be the case that λ is real. All other roots of (4.3) must have real part strictly smaller than λ since all eigenvalues of $T(r)$ have modulus strictly smaller than ρ. Thus $\lambda = s(L)$ and since $y(t) = e^{\lambda t}$ is a solution of (4.1) it follows that $y_r = \psi$ and $0 \ll_\mu y_r$. By the definition of the strong ordering, this implies that the derivative of $e^{\mu s}e^{\lambda(r+s)}$ is positive on $[-r, 0]$. Hence, $\lambda + \mu > 0$ so $s(L) > -\mu$. $\square$

The ordinary differential equation obtained from (3.1) by ignoring the delay is given by
$$x' = F(x), \quad F(x) = f(\hat{x}). \tag{4.4}$$
Equations (4.4) and (3.1) share the same equilibria. We now show that if v is an equilibrium such that (4.2) holds and if v is asymptotically stable (unstable) in the linear approximation for (4.4), then it is asymptotically stable (unstable) for (4.1).

COROLLARY 4.2. *Let v be an equilibrium of (3.1) and suppose that (4.2) holds. If $F'(v) > 0$ then $s(L) > 0$ and if $F'(v) < 0$ then $s(L) < 0$.*

PROOF. $F'(v) = df(\hat{v})(\hat{1}) = Le_0$, where we use the notation e_λ defined in (4.3). If $F'(v) > 0$, then $Le_0 - 0 > 0$. For very large real λ, $Le_\lambda - \lambda < 0$ since $Le_\lambda \to \nu(\{0\})$ as $\lambda \to \infty$ where ν is the Borel measure representing L. The intermediate value theorem implies the existence of a positive root of (4.3). Consequently, $s(L) > 0$.

If $F'(v) < 0$, then $Le_0 - 0 < 0$. On the other hand, since $0 <_\mu e_{-\mu}$, (4.2) implies that $Le_{-\mu} - (-\mu) > 0$. Again, the intermediate value theorem implies the existence of a root, λ_0, of (4.3) satisfying $-\mu < \lambda_0 < 0$. Therefore, $0 \ll_\mu e_{\lambda_0}$ and e_{λ_0} is an eigenvector of $T_L(r)$ corresponding to the eigenvalue $e^{\lambda_0 r}$. By the Krein-Rutman theorem, there is only one eigenvector in K_μ, up to scalar multiple. Consequently, $\lambda_0 = s(L)$ and so $s(L) < 0$. $\square$

Again we see that for monotone dynamical systems, the stability of an equilibrium is always governed by a real dominant eigenvalue. Essentially, this can be viewed as a necessary condition for a system to be monotone. While a Hopf bifurcation to a small amplitude periodic solution is possible in a monotone system, it can never be attracting by Theorem 1.2.2.

5. A Model of an Adult Fly Population

In this section, we illustrate the results of the previous sections by examining an equation derived by Gurney et al(1980) as a model of the population dynamics of a species of fly. Although the focus of the investigation of Gurney et al was the oscillatory behavior of the adult population in time, our study will focus on the case where convergence to equilibrium occurs. The equation for the adult population size N as a function of time s is given by

$$\frac{dN}{ds}(s) = -\delta N(s) + pN(s-r)\exp(-qN(s-r)). \tag{5.1}$$

It is convenient to scale (5.1) by letting

$$x = qN, \ s = rt, \ a = \delta r, \ b = pr$$

which leads to the equation with unit delay

$$x'(t) = -ax(t) + bx(t-1)\exp(-x(t-1)). \tag{5.2}$$

Obviously, only nonnegative solutions of (5.2) are of interest. Theorem 5.2.1 insures that all solutions corresponding to nonnegative initial data are nonnegative on their domain of existence. Hereafter, by a solution of (5.2) we always mean a nonnegative solution.

It is of interest to compare the behavior of solutions of (5.2) with the ordinary differential equation obtained by ignoring the delay. This equation, given by

$$u' = -au + bu\exp(-u), \tag{5.3}$$

has the same equilibria as (5.2). If $b/a < 1$, (5.3) has only the trivial equilibrium $u = 0$ which is asymptotically stable and attracts all other solutions. If $b/a > 1$ then, in addition to the trivial equilibrium, which is unstable, there is a positive equilibrium

$$v = \log(b/a), \tag{5.4}$$

which is asymptotically stable and attracts all nontrivial solutions. Under additional assumptions, we will show that the same result holds for (5.2). However, because a Hopf bifurcation from the equilibrium (5.4) to small amplitude periodic solutions can occur for (5.2), as noted in Gurney et al(1980), some restrictions on the parameters are necessary for our results to hold and it will be interesting to compare this parameter region with the region where (5.4) is asymptotically stable.

The nonlinear recruitment rate $h(y) = by\exp(-y)$ is a hump-shaped function, reaching its maximum value be^{-1} at $y = 1$ before decreasing to zero, and this lack of monotonicity might suggest that the results of Chapter 5 do not apply. In fact, they do apply provided that whenever v in (5.4) is positive, then it lies in the region $0 \leq v \leq 1$ where h is monotone increasing. In other words, when $b/a \leq e$. In that case we have the following result.

THEOREM 5.1. *If $b/a < 1$ then the trivial solution of (5.2) attracts all other solutions. If $1 < b/a \leq e$, then the nontrivial equilibrium (5.4) attracts all nontrivial solutions of (5.2).*

PROOF. The plan is to first show that the result holds with the restriction that the initial data belong to the order interval $I = [\hat{0}, \hat{1}]$ and then to show that all solutions corresponding to nonnegative initial data eventually enter this order interval. By Remark 2.1 of the previous chapter, to show that I is positively invariant for (5.2) we must show that if $\phi \in I$ and $\phi(0) = 1$, then $-a + h(\phi(-1)) \leq 0$. As $h(\phi(-1)) \leq h(1) = be^{-1}$ and $b/a \leq e$, the above inequality holds and I is positively invariant. Since h is increasing on the interval $[0,1]$, (5.2) satisfies the quasimonotone condition (Q) on I. The proof of Proposition 5.4.2 shows that every solution of (5.2) corresponding to initial data $\phi \in I$ converges to the trivial equilibrium if $b/a < 1$. If $1 < b/a \leq e$ then the same argument shows that each such solution converges to one of the two equilibria. We must show that all but the trivial solution converges to (5.4) in this case. As $b > a$, choose $\epsilon_0 > 0$ such that $-a + be^{-\epsilon_0} > 0$. If $0 < \epsilon \leq \epsilon_0$ and if $f(x_t)$ denotes the right side of (5.2), then $f(\hat{\epsilon}) > 0$. Consequently, Corollary 5.2.2 implies that $x(t, \hat{\epsilon})$ is nondecreasing and converges to (5.4) as $t \to \infty$. If $\phi \in I$ and $0 \ll \phi$, then $\hat{\epsilon} \leq \phi$ for some $\epsilon < \epsilon_0$ so $x(t, \hat{\epsilon}) \leq x(t, \phi)$ for all $t \geq 0$ by monotonicity. Therefore, $x(t, \phi)$ converges to (5.4). Now it is easy to see that if $\phi > 0$, then $x(t, \phi) > 0$ for all $t > 1$ so the above comparison argument can be applied to $x(t, x_2(\phi))$ to show that $x(t, \phi)$ converges to (5.4). This establishes that the assertions of the theorem hold for initial data belonging to I.

Now we show that all solutions eventually enter I. To this end note that the same arguments used to show that I is positively invariant also show that $I_L = [\hat{0}, \hat{L}]$ is positively invariant for all $L \geq be^{-1}/a$. Therefore, if $\phi \in I_L$ for some $L \geq 1$ then $x_t(\phi) \in I_L$ for all $t \geq 0$. The omega limit set $\omega(\phi)$ is nonempty, compact and invariant for (5.2). Let $\psi \in \omega(\phi)$ be such that $\psi(0) = \max\{\xi(0) : \xi \in \omega(\phi)\}$. Since $\omega(\phi)$ is invariant, $|\xi| \leq \psi(0)$ for all $\xi \in \omega(\phi)$ and there exists $\eta \in \omega(\phi)$ such that $x_1(\eta) = \psi$. Suppose that $\psi(0) > 1$. Since $\psi(s) \leq \psi(0)$ for all $s \in [-1, 0]$, a calculation shows that $x'(0, \psi) = x'(1, \eta) < 0$. But this means that $x(t, \eta) > \psi(0)$ for some $t < 1$ in contradiction to the maximality of $\psi(0)$. Therefore, $\psi(0) \leq 1$ and $\omega(\phi) \subset I$. The lemma below shows that $0 \notin \omega(\phi)$ for all $\phi > 0$. By the invariance of $\omega(\phi)$ and the fact that every nontrivial solution corresponding to initial data in I converges to (5.4), it follows that $\omega(\phi) = \{\hat{v}\}$ where v is given in (5.4). □

LEMMA 5.2. *If $b/a > 1$ then $\hat{0} \notin \omega(\phi)$ for all $\phi > 0$.*

PROOF. As noted in the previous proof, if $\phi > 0$ then $x(t) = x(t, \phi) > 0$ for all $t > 1$. We first note that $x(t)$ does not converge to zero as t increases to infinity by the comparison argument of the first paragraph of the previous proof. Consequently, $\limsup_{t \to \infty} x(t) > 0$. If the result is false, then $\liminf_{t \to \infty} x(t) = 0$ and we may choose a sequence $t_n \to \infty$ such that $x(t_n) \to 0$ as $n \to \infty$, $x'(t_n) = 0$ and $x(t_n - 1) > x(t_n)$. By (5.2),

$$0 = -ax(t_n) + h(x(t_n - 1)) > -ax(t_n) + h(x(t_n)) > 0,$$

a contradiction which completes the proof. □

The techniques of the present chapter are used in the next result to extend the last assertion of Theorem 5.1 to a larger parameter region. Note that the first inequality of the hypothesis below just says that the equilibrium (5.4) lies to the right of $y = 1$ where h reaches its maximum.

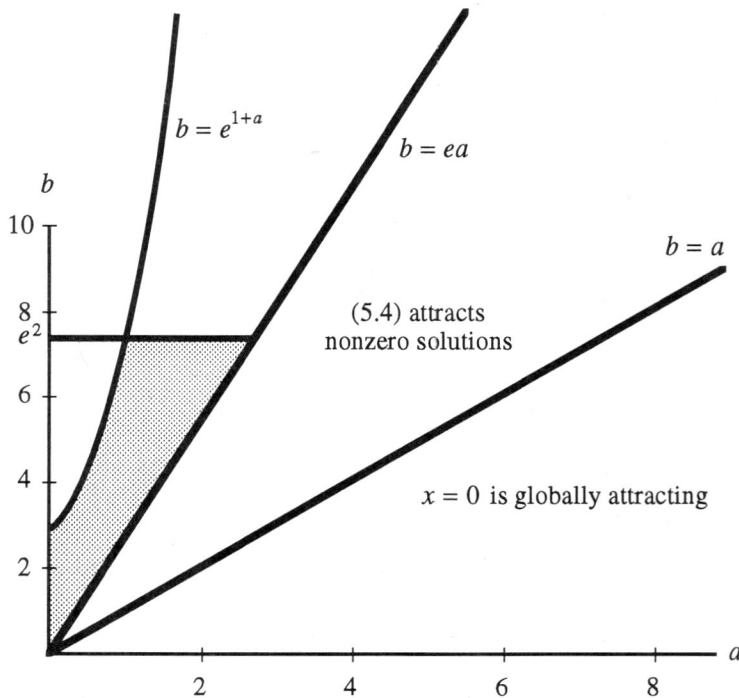

FIGURE 5.1. Hypotheses of Theorem 5.3 hold in shaded region.

THEOREM 5.3. *If $b/a > e$ and $b < e^2$ and $b < e^{1+a}$, then equilibrium (5.4) attracts all nontrivial solutions of (5.2).*

PROOF. (5.2) has the form of (0.1) so by Remark 2.2, (SM_μ) holds for some $\mu \geq 0$ provided condition (iv) holds where $L_1 = -a$ and $L_2 = -be^{-2}$. The second and third inequalities of the hypothesis of the theorem imply that (iv) of Remark 2.2 holds on the domain $D = C_+$. The proof of Theorem 5.1 showed the positive invariance of each order interval $I_L = [\hat{0}, \hat{L}]$ for $L \geq be^{-1}/a$. This is easily seen to imply that (T) holds and therefore, by Theorem 3.1, all solutions of (5.2) converge to one of the two equilibria. However, Lemma 5.2 shows that no nontrivial solution is attracted to the trivial equilibrium. Therefore, all nontrivial solutions converge to (5.4). □

Figure 5.1 depicts the region in parameter space where Theorems 5.1 and 5.3 apply.

Although the stability of the equilibria of (5.2) did not play an essential role in either result, it is worth mentioning that the trivial equilibrium of (5.2) is asymptotically stable when $b < a$ and unstable when $b > a$; the equilibrium (5.4) is asymptotically stable under the hypotheses of both Theorems 5.1 and 5.2 when $b > a$. These assertions follow from Corollary 5.5.2 and Corollary 4.2. However, (5.4) is not always stable. The variational equation about (5.4) is given by

$$z'(t) = -Az(t) - Bz(t-1),$$

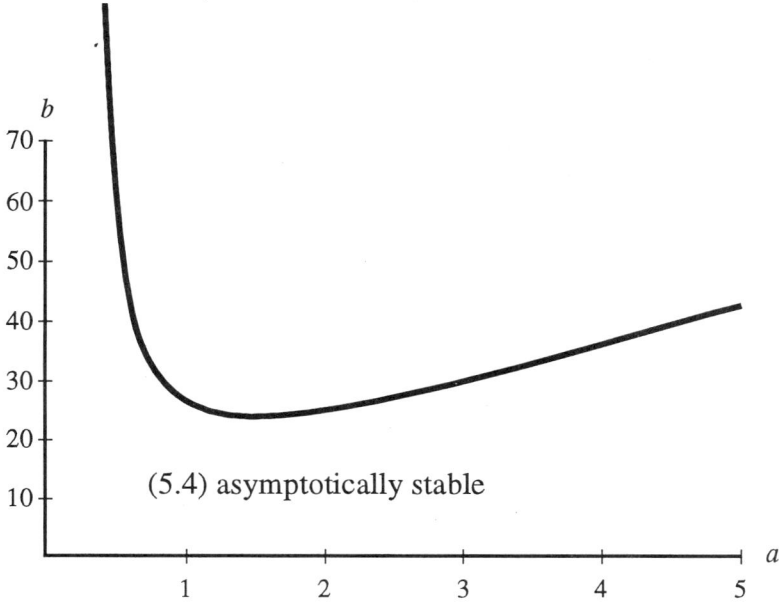

FIGURE 5.2. The neutral stability curve for (5.4).

where
$$A = a \quad \text{and} \quad B = a(\log(b/a) - 1). \tag{5.5}$$

This leads to the characteristic equation

$$(\lambda + A)e^\lambda + B = 0 \tag{5.6}.$$

Necessary and sufficient conditions for every root of (5.6) to have negative real part, due to Hayes(1950), are described in Hale(1977). These are that $A > -1$, $A + B > 0$ and $B < \eta \sin(\eta) - A \cos(\eta)$ where η is the root of $\eta = -A \tan(\eta)$, $0 < \eta < \pi$ if $A \neq 0$ and $\eta = \pi/2$ if $A = 0$. The first two conditions are trivially satisfied in the case that (5.5) holds with $a > 0$ and $b > a$. In fact, since $A > 0$, we see that η can be restricted to the interval $\pi/2 < \eta < \pi$. Therefore, the neutral stability curve is given by

$$A = -\eta \cot(\eta)$$
$$B = \eta \sin(\eta) + \eta \cot(\eta) \cos(\eta),$$

where η belongs to the above interval. In terms of a and b, the neutral stability curve is given by

$$a = -\eta \cot(\eta)$$
$$b = -\eta \cot(\eta) \exp(1 - \sec(\eta)).$$

Equilibrium (5.4) is asymptotically stable for parameters (a, b) below this curve. **Figure 5.2** depicts the neutral stability curve. See Gurney et al(1980) for a similar plot. It can be shown that a Hopf bifurcation occurs along this curve. Numerical simulations in Gurney et al(1980) suggest that stable sustained oscillations exist for some parameters situated above the curve.

Figure 5.1 shows the region where monotonicity methods are successful and Figure 5.2 shows the region of asymptotic stability for the equilibrium (5.4) (but not necessarily the region where it is globally attracting). It would be interesting to know if (5.4) is globally attracting in the parameter region where it is asymptotically stable.

6. Remarks and Discussion

The material for this chapter was taken from Smith and Thieme(1990b,1991b), but curiously, the exponential ordering was used earlier by Hadeler and Tomiuk (1977) to show the existence of nontrivial periodic solutions of delay differential equations using fixed point methods. We use it to exclude periodic solutions and to show that most solutions converge to equilibrium.

The novel feature encountered in this chapter is the use of a cone which has empty interior in the natural state space. Because the cone $\tilde{K}_\mu$ has empty interior in C, the semiflow generated by (1.1) cannot be eventually strongly monotone. It might be thought that by choosing C_L as state space instead of C then all our problems would go away since K_μ has nonempty interior in C_L. However, (1.1) does not generate a semiflow in the space C_L, in general, since the map from $\mathbb{R}_+$ to C_L given by $t \to x_t(\phi)$ is not continuous at $t = 0$ for some $\phi \in C_L$. Consider, for example the equation $x' = 1$ with $\phi = 0$. Then $|\phi|_L = 0$ but $|x_t(\phi)|_L = t + 1$ for $t > 0$. Therefore, working in the space C_L would create difficulties and even if we were successful in surmounting them, it would not be easy to exploit the information gained about the dynamical system in C_L to draw conclusions in the natural space C. The advantage of the SOP condition is apparent in this chapter.

The presentation here may have suggested that the space C_L was pulled out of a hat. In fact, as noted in Smith and Thieme(1990b), it arises in a natural way by a standard method. If we set $\xi(s) = \exp((1 - \mu)s)$, then $\xi \in \tilde{K}_\mu$ and we can define the space

$$X = \{\phi \in C : \exists \beta \geq 0, -\beta\xi \leq_\mu \phi \leq_\mu \beta\xi\}.$$

It may be shown that

$$|\phi|_\mu = \inf\{\beta \geq 0 : -\beta\xi \leq_\mu \phi \leq_\mu \beta\xi\}$$

defines a norm on X which makes it a Banach space and furthermore $K_\mu = \tilde{K}_\mu \cap X$ is a cone with nonempty interior in X. It turns out this space is just C_L. In fact, C_L is a Banach lattice with the ordering induced by K_μ. See Amann(1976) for a general discussion.

The exponential ordering idea is extended to systems of delay differential equations in Smith and Thieme(1991b). The key is to replace the scalar $-\mu$ by a

6. REMARKS AND DISCUSSION

quasipositive matrix B (see Chapter 4, sec.3) in defining a cone:

$$K_B = \{\phi \in C : \phi \geq 0 \text{ and } \phi(t) \geq \exp(B(t-s))\phi(s), \ -r \leq s \leq t \leq 0\}.$$

A smooth function belongs to K_B if and only if

$$\phi' \geq B\phi \quad \text{and} \quad \phi \geq 0 \text{ on } [-r, 0].$$

We write $\phi \leq_B \psi$ when $\psi - \phi \in K_B$. The condition (M_μ) is replaced by

(M_B) Whenever $\phi, \psi \in C$ and $\phi \leq_B \psi$ then $f(\psi) - f(\phi) \geq B[\psi(0) - \phi(0)]$.

In order to give an idea of the kind of results that can be expected for systems of delay differential equations using this ordering, consider the system (0.1) where x and g are vectors in $\mathbb{R}^n$ and g is continuously differentiable. Then (M_B) is satisfied provided that

$$\frac{\partial g}{\partial x}(x, y) \geq B$$

$$\left(\frac{\partial g}{\partial x} - B\right)e^{Br} + \frac{\partial g}{\partial y} \geq 0,$$

holds for all (x, y). In particular, the partial derivative of g with respect to x must be quasipositive and since $e^{Br} > 0$, the requirement that the partial derivative of g with respect to y be nonnegative (in all entries), as required in the previous chapter, can be relaxed. See Smith and Thieme(1991b) for a more thorough treatment and some examples.

The results of section 4 on stability of equilibria are not the best possible. In Smith and Thieme(1990b) it is shown that $s(L)$ is a root of the characteristic equation if (4.2) is relaxed to

$$L\phi + \mu\phi(0) \geq 0, \quad \text{whenever } 0 \leq_\mu \phi.$$

More general results in this direction for general positive semigroups of operators can be found in Kerscher and Nagle(1984). See also the remarks and discussion section of the previous chapter.

As noted in section 5, the model equation (5.1) is due to Gurney et al(1980). A nice discussion of the biology and some numerical simulations showing sustained oscillations for certain parameter regions are given in Nisbet and Gurney(1982).

CHAPTER 7

Quasimonotone Systems of Parabolic Equations

0. Introduction

Monotonicity methods have traditionally played a larger role in the theory of partial differential equations than for ordinary or delay differential equations. Maximum principles lie at the core of the theory of elliptic and parabolic equations. In the present chapter, these well-developed methods facilitate the application of monotone systems theory to initial-boundary value problems associated with systems of nonlinear parabolic partial differential equations. In the applied literature, these systems are often referred to as reaction-diffusion systems since they arise naturally as models of reacting and diffusing chemical species.

We consider a model reaction-diffusion system and sketch an outline of how it may be viewed as a semiflow on a suitable Banach space of functions on a spatial domain. While there are a variety of suitable state spaces from which to choose, a natural choice for a brief treatment is a subspace of the continuous functions since this allows us to avoid a lengthy and technical discussion of function spaces and besides, it is a familiar one by now. Once we have set up a semiflow, the usual problems arise: when is a subset positively invariant, when can comparisons be made between solutions of possibly different systems, under what circumstances is the semiflow monotone, strongly monotone or strongly order preserving. For parabolic partial differential equations, these questions require the use of various maximum principles and therefore, an entire section is devoted to maximum principles and there implications.

The main result of this chapter will not be a surprise. We show that the generic solution of a cooperative and irreducible reaction-diffusion system converges to an equilibrium, that is, to a solution of the associated elliptic boundary value problem. The stability of an equilibrium then becomes an especially important issue. As usual, the Krein-Rutman theorem can be used to show that stability is determined by a real eigenvalue and the corresponding eigenfunction has certain positivity properties depending on the boundary conditions. Robin, Neumann and Dirichlet boundary conditions are considered.

We begin by describing the model problem in the next section. It can most easily be viewed as a semi-dynamical system by considering so-called mild solutions of a corresponding abstract ordinary differential equation in a suitable Banach space. However, enough smoothness is hypothesized that these mild solutions are, in fact, classical solutions of the partial differential equation. The treatment draws more heavily from semigroup methods than previous chapters.

As the reader will know, issues of smoothness of various coefficients and terms appearing in differential equations greatly affect questions of existence, uniqueness and regularity of solutions and this is especially so in the theory of partial differential equations. Our philosophy is to make appropriate smoothness hypotheses for those terms appearing in our model system, once and for all, such that they are sufficient for all the results in this chapter. In this way, we avoid the distraction of continually tinkering with them. As a consequence however, many of these results hold under weaker smoothness conditions. The reader may wish to keep in mind, when smoothness conditions are made, that ultimately, we require the parabolic system to generate a C^1 semiflow on a suitable state space. This will ensure, among other things, that a linearized stability analysis can be rigorously justified.

1. Parabolic Systems: The Basic Setup

We begin by describing the model system and how we view it as a semi-dynamical system. Let Ω be a bounded, open and connected subset of $\mathbb{R}^N$ with a smooth boundary, $\partial\Omega$. By smooth, we mean that the $\partial\Omega$ can be locally described as the level set of a function of class C^4 with nonvanishing gradient. Let Δ be the Laplacian operator and ∇ be the gradient operator on Ω and consider the reaction diffusion system

$$\frac{\partial u_i}{\partial t}(x,t) = d_i \Delta u_i(x,t) + F_i(x, u(x,t)), \quad t > 0, \ x \in \Omega,$$
$$0 = \alpha_i(x) u_i(x,t) + \delta_i \frac{\partial u_i}{\partial \nu}(x,t), \quad t > 0, \ x \in \partial\Omega, \tag{1.1}$$
$$u_i(x,0) = \phi_i(x), \quad x \in \Omega,$$

where $i = 1, 2, \ldots, n$ and $d_i > 0$ are constants. The boundary conditions are assumed to be either Dirichlet conditions:

$$\delta_i = 0 \quad \text{and} \quad \alpha_i(x) \equiv 1,$$

or Neumann or Robin conditions:

$$\delta_i = 1 \quad \text{and} \quad \alpha_i(x) \geq 0.$$

If α_i vanishes identically, the boundary condition is of Neumann type, otherwise it is of Robin, or mixed type. It is assumed that $\alpha_i(x)$ is twice continuously differentiable on $\bar{\Omega}$. The vector field $\nu = \nu(x)$ is the outer unit normal to $\partial\Omega$ at $x \in \partial\Omega$ and $\frac{\partial}{\partial \nu} = \nu \cdot \nabla$ denotes differentiation in the direction of the outward normal to Ω. If $\delta_i = 1$, then the initial data ϕ_i is assumed to belong to the space $C(\bar{\Omega}) = C(\bar{\Omega}, \mathbb{R})$ with the usual supremum norm. If $\delta_i = 0$, then $\phi_i \in C_0(\bar{\Omega})$, the subspace of $C(\bar{\Omega})$ of functions vanishing on $\partial\Omega$. For simplicity in the statement of results, we assume that either $\delta_i = 0$ for all i or for no i. That is, we assume that either the Dirichlet boundary conditions are used for all components or the Robin or Neumann conditions are used for all components of (1.1). We stress that this restriction is not necessary and that its sole purpose is to allow a more concise statement of results.

Let $\phi = (\phi_1, \phi_2, \ldots, \phi_n)$ and similarly define F and u. Then, we can express (1.1) in the more succinct form

$$\frac{\partial u}{\partial t}(x,t) = D\Delta u(x,t) + F(x, u(x,t)), \quad t > 0, \ x \in \Omega$$
$$0 = \alpha(x) u(x,t) + \beta \frac{\partial u}{\partial \nu}(x,t), \quad t > 0, \ x \in \partial\Omega$$
$$u(x,0) = \phi(x), \quad x \in \Omega,$$

1. PARABOLIC SYSTEMS: THE BASIC SETUP

in which D is the diagonal matrix of diffusivities d_i, $\beta = I$ for Robin or Neumann boundary conditions and $\beta = 0$ for Dirichlet conditions and $\alpha(x)$ is the diagonal matrix with diagonal entries $\alpha_i(x)$.

Various generalizations of (1.1) could be considered. For example, all of the results of this chapter hold if the operators $d_i \Delta$ are replaced by any second order differential operator

$$A_i = \sum_{k,j=1}^{N} a_{kj}^i(x) D_k D_j + \sum_{k=1}^{N} a_k^i(x) D_k$$

where $a_{kj}^i(x) = a_{jk}^i(x)$, $D_k = \frac{\partial}{\partial x_k}$. The coefficients a_{kj}^i and a_k^i are assumed to be twice continuously differentiable on $\bar{\Omega}$. We also assume that A_i is *uniformly elliptic* in Ω in the sense that there exists a positive constant μ such that for all $\xi \in \mathbb{R}^N$ we have

$$\sum_{k,j=1}^{N} a_{kj}^i(x) \xi_k \xi_j \geq \mu |\xi|^2, \quad x \in \Omega.$$

Our aim is to show that (1.1) generates a dynamical system on a suitable state space. We begin by expressing it as an abstract ordinary differential equation in the Banach space $X = \prod_1^n X_i$ where, for Robin or Neumann boundary conditions, $X_i = C(\bar{\Omega})$, or if Dirichlet boundary conditions are used, $X_i = C_0(\bar{\Omega})$. In either case, the norm on X is defined by $|\phi| = \sum |\phi_i|$. For each i, let A_i^0 be the differential operator

$$A_i^0 u_i = d_i \Delta u_i$$

defined on the domain $D(A_i^0) \subset X_i$ given by

$$D(A_i^0) = \{u_i \in C^2(\Omega) \cap C^1(\bar{\Omega}) : A_i^0 u_i \in C(\bar{\Omega}), \; \alpha_i(x) u_i(x) + \delta_i \frac{\partial u_i}{\partial \nu}(x) = 0, \; x \in \partial\Omega\},$$

in the case of Robin or Neumann boundary conditions, and by

$$D(A_i^0) = \{u_i \in C_0(\bar{\Omega}) \cap C^2(\Omega) : A_i^0 u_i \in C_0(\bar{\Omega})\}$$

in the case of Dirichlet boundary conditions. The closure A_i of A_i^0 in X_i generates an analytic semigroup of bounded linear operators $T_i(t)$ for $t \geq 0$ such that $u_i(t) = T_i(t)\phi_i$ is the solution of the abstract linear differential equation in X_i given by

$$u_i'(t) = A_i u_i(t), \quad u_i(0) = \phi_i \in D(A_i).$$

The domain $D(A_i)$ is characterized implicitly by

$$D(A_i) = \{\psi \in X_i : \lim_{t \to 0+} t^{-1}(T_i(t) - I)\psi \text{ exists}\}$$

and the limit is precisely $A_i \psi$. We do not need an explicit characterization of $D(A_i)$ at this time but see the proof of Theorem 3.1.

The defining properties of the semigroup $T_i(t)$ are
(1) $T_i(0) = I$, (the identity map on X_i)
(2) $T_i(t) T_i(s) = T_i(t+s)$, $t, s \geq 0$,
(3) $t \to T_i(t)\psi$ is a continuous map from $[0, \infty)$ into X_i for all $\psi \in X_i$.
An additional property of the semigroup that will be important later is:
(4) For each $t > 0$, $T_i(t) : X_i \to X_i$ is a compact operator.
See the estimate (4.3) below, which leads immediately to (4).

In the language of partial differential equations,
$$u_i(x,t) = [T_i(t)\phi_i](x)$$
is a classical solution of the linear initial boundary value problem
$$\frac{\partial u_i}{\partial t}(x,t) = d_i \Delta u_i(x,t), \quad t > 0, \ x \in \Omega,$$
$$0 = \alpha_i(x) u_i(x,t) + \delta_i \frac{\partial u_i}{\partial \nu}(x,t), \quad t > 0, \ x \in \partial\Omega, \quad (1.2)$$
$$u_i(x,0) = \phi_i(x), \quad x \in \Omega.$$

That is, the derivatives $\frac{\partial u_i}{\partial t}, \frac{\partial u_i}{\partial x_j}, \frac{\partial^2 u_i}{\partial x_j \partial x_k}$ exist and are continuous for $(x,t) \in \Omega \times (0,\infty)$, the derivatives $\frac{\partial u_i}{\partial x_j}$ are continuous on $\bar{\Omega} \times (0,\infty)$ and u_i satisfies the equation and the boundary conditions. Furthermore, u_i is continuous on $\bar{\Omega} \times [0,\infty)$ and $u_i(x,0) = \phi_i(x)$ for $x \in \Omega$.

Let $T(t) : X \to X$ be defined by $T(t) = \prod_1^n T_i(t)$. Then $T(t)$ is a semigroup of operators on X generated by the operator $A = \prod_1^n A_i$ defined on $D(A) = \prod_1^n D(A_i)$ and $u(x,t) = [T(t)\phi](x)$ is the solution of the linear system
$$\frac{\partial u}{\partial t}(x,t) = D\Delta u(x,t), \quad t > 0, \ x \in \Omega$$
$$0 = \alpha(x) u(x,t) + \beta \frac{\partial u}{\partial \nu}(x,t), \quad t > 0, \ x \in \partial\Omega$$
$$u(x,0) = \phi(x), \quad x \in \Omega.$$

Concerning the nonlinear term $F : \bar{\Omega} \times \mathbb{R}^n \to \mathbb{R}^n$ in (1.1), we assume hereafter that it is twice continuously differentiable in (x,u). In the case of Dirichlet boundary conditions, we also assume that
$$F_i(x,u) = 0 \text{ whenever } x \in \partial\Omega \text{ and } u_i = 0. \quad (1.3)$$

(1.3) is not needed for Robin or Neumann boundary conditions. For the applications we have in mind, the restriction (1.3) is not a serious one. Typically, the u_i are nonnegative variables (e.g. concentrations or densities) and $F_i = 0$ when $u_i = 0$. With these assumptions, the map f defined by
$$[f(\phi)](x) = F(x, \phi(x)),$$
maps X into itself. Equation (1.1) can then be viewed as the abstract ordinary differential equation in X given by
$$u'(t) = Au(t) + f(u(t)), \quad u(0) = \phi. \quad (1.4)$$

While a solution $u(t)$ of (1.4) can be obtained under the restriction that $\phi \in D(A)$, a so-called *mild solution* can be obtained for every $\phi \in X$ by requiring only that $u(t)$ is a continuous solution of the variation of constants expression
$$u(t) = T(t)\phi + \int_0^t T(t-s) f(u(s)) ds. \quad (1.5)$$

More precisely, $u : [0,\tau) \to X$ is a mild solution of (1.5) if it is continuous and satisfies (1.5) on $[0,\tau)$. It turns out, as we shall see in Theorem 3.1 below, that with the smoothness assumptions above, a unique mild solution of (1.5) exists and we can define a semiflow on the space X by

$$\Phi_t(\phi) = u(t, \phi)$$

where $u(t, \phi)$ is the solution of (1.5) satisfying $u(0, \phi) = \phi$. Furthermore, the mild solution of (1.5) is also a classical solution of (1.1). These issues are dealt with in section 3.

As expected, the equilibria will play a central role in the dynamics of (1.1). An equilibrium solution of (1.1) is a time-independent (classical) solution and therefore can be viewed as a function $v \in C^2(\Omega) \cap C^1(\bar{\Omega})$ satisfying

$$\begin{aligned} 0 &= D\Delta v(x) + F(x, v(x)), \ x \in \Omega \\ 0 &= \alpha(x)v(x) + \beta \frac{\partial v}{\partial \nu}(x), \ x \in \partial \Omega. \end{aligned} \quad (1.6)$$

Having outlined the model problem and sketched our point of view on solutions, the next task is to introduce the primary tools, namely, parabolic maximum principles.

2. Maximum Principles

Maximum principles play a central role in the theory of parabolic partial differential equations. They provide the fundamental tools for establishing the existence of positively invariant sets, comparisons between solutions of different parabolic equations and monotonicity of the solution operator. The classic reference is the book by Protter and Weinberger(1967) and we follow it here.

The aim of this section is to introduce various maximum principles which will be useful for obtaining results of the kind just mentioned. We also apply these results to show that the semigroup $T_i(t)$, generated by (1.2), is a positive semigroup (strongly positive in the case of Robin or Neumann boundary conditions) and that certain order intervals are positively invariant.

Consider the general second order parabolic operator

$$Au = \sum_{i,j=1}^{N} a_{ij}(x,t)D_i D_j u + \sum_{i=1}^{N} a_i(x,t)D_i u - \frac{\partial u}{\partial t},$$

for the function $u(x,t)$ with (x,t) belonging to a cylinder $\Omega \times (0,T)$ for some $T > 0$. The symbol D_i stands for the partial derivative $\frac{\partial}{\partial x_i}$. The coefficients $a_{ij}(x,t) = a_{ji}(x,t)$ and $a_i(x,t)$ are assumed to be continuous on $\bar{\Omega} \times [0,T]$. A is called *uniformly parabolic* on $\Omega \times (0,T)$ if there exists a positive constant μ such that for any $\xi \in \mathbb{R}^N$

$$\sum_{i,j=1}^{N} a_{ij}(x,t)\xi_i \xi_j \geq \mu |\xi|^2, \quad (x,t) \in \Omega \times (0,T).$$

Throughout this section, let

$$Q = \partial \Omega \times (0,T).$$

The Strong Parabolic Maximum Principle, stated below, says that a nonconstant solution of a parabolic partial differential inequality on a cylindrical domain assumes its maximum on a restricted part of the boundary of the domain.

THEOREM 2.1. (*Strong Parabolic Maximum Principle*) *Let A be uniformly parabolic in $\Omega \times (0,T)$ and let the function $u(x,t)$ be continuous on $\bar{\Omega} \times [0,T]$ and the derivatives $\frac{\partial u}{\partial x_i}, \frac{\partial^2 u}{\partial x_i \partial x_j}, \frac{\partial u}{\partial t}$ exist and are continuous on $\Omega \times (0,T]$. Finally, assume that u satisfies*

$$Au(x,t) \geq 0, \quad (x,t) \in \Omega \times (0,T]. \tag{2.1}$$

If u attains its maximum on $\bar{\Omega} \times [0,T]$ at a point $p = (\bar{x}, \bar{t}) \in \Omega \times (0,T]$, then $u(x,t) = u(p)$ for all $(x,t) \in \bar{\Omega} \times [0,T]$ such that $t \leq \bar{t}$. If the inequality (2.1) is reversed and if u attains its minimum in $\bar{\Omega} \times [0,T]$ at a point $p = (\bar{x}, \bar{t}) \in \Omega \times (0,T]$, then the same conclusion holds.

The next result says that the normal derivative of the solution of (2.1) must be positive at a boundary point where the maximum of the solution is attained.

THEOREM 2.2. *Let A be uniformly parabolic in $\Omega \times (0,T)$ and let $u(x,t)$ satisfy the hypotheses of Theorem 2.1. In addition, suppose that the first partial derivatives of u with respect to the x_i exist and are continuous on $(\Omega \times (0,T]) \cup Q$. If u attains its maximum on $\bar{\Omega} \times [0,T]$ at a point $p = (\bar{x}, \bar{t}) \in Q$ and $u(x,t) < u(p)$ for $(x,t) \in \Omega \times (0,T)$, then $\frac{\partial u}{\partial \nu}(p) > 0$. A similar statement holds in the case that the inequality is reversed in (2.1): if u attains its minimum on $\bar{\Omega} \times [0,T]$ at a point $p = (\bar{x}, \bar{t}) \in Q$ and $u(x,t) > u(p)$ for $(x,t) \in \Omega \times (0,T)$, then the opposite inequality holds for the normal derivative of u at p.*

The maximum principles can be used to show that the $T_i(t)$ are positive linear operators. Let X_i^+ be the cone of nonnegative functions in X_i and $\leq$ be the corresponding partial order relation on X_i. As usual, $\phi_i < \psi_i$ means $\phi_i \leq \psi_i$ but $\phi_i \neq \psi_i$. If $X_i = C(\bar{\Omega})$, then X_i^+ has nonempty interior in X_i consisting of the functions which are positive at all points of $\bar{\Omega}$. In this case, the strong order relation $\ll$ is defined and $\phi_i \ll \psi_i$ holds if and only if $\phi_i(x) < \psi_i(x)$ holds for all $x \in \bar{\Omega}$. The next result says that $T_i(t)$ is strongly positive in this case. However, if $X_i = C_0(\bar{\Omega})$ then X_i^+ has empty interior in X_i so the strong order relation is not defined.

COROLLARY 2.3. *The semigroup $T_i(t)$ is positive. More precisely,*

$$T_i(t) X_i^+ \subset X_i^+.$$

Furthermore, if $\phi_i > 0$ and if $u_i(x,t) = [T_i(t)\phi_i](x)$, then $u_i(x,t) > 0$ holds for all $t > 0$ and $x \in \bar{\Omega}$ in the case of Neumann or Robin boundary conditions and for all $t > 0$ and $x \in \Omega$ in case of Dirichlet boundary conditions. In addition, $\frac{\partial u}{\partial \nu}(x,t) < 0$ for all $(x,t) \in \partial\Omega \times (0,\infty)$ in the case of Dirichlet boundary conditions.

PROOF. Observe that (1.2) is uniformly parabolic. Let $u(x,t) = u_i(x,t) = [T_i(t)\phi_i)](x)$ for some $\phi_i \in X_i$ satisfying $\phi_i \geq 0$. First consider the case of Dirichlet boundary conditions. If the result were false, then $u(x^*, t^*) < 0$ for some point $q = (x^*, t^*)$ with $t^* > 0$. Choose $T > t^*$ and suppose that u attains its minimum value on $\bar{\Omega} \times [0,T]$ at $p = (\bar{x}, \bar{t})$. Obviously, $u(p) < 0$ and $0 < \bar{t} \leq T$ and $\bar{x} \in \Omega$, the latter since $u = 0$ on $\partial\Omega$. According to Theorem 2.1, $u(\bar{x}, 0) = \phi_i(\bar{x}) = u(p)$, which contradicts that $\phi_i(\bar{x}) \geq 0$. Therefore, $u(x,t) \geq 0$ for all $(x,t) \in \Omega \times [0,\infty)$. Suppose now that $\phi_i(\bar{x}) > 0$ for some $\bar{x} \in \Omega$ and fix $T > 0$. If $u(x^*, t^*) = 0$ for some $(x^*, t^*) \in \Omega \times (0,T)$, then, by the argument above, u attains its minimum on $\bar{\Omega} \times [0,T]$ at (x^*, t^*) so by Theorem 2.1, $u(\bar{x}, 0) = \phi_i(\bar{x}) = 0$. This contradiction proves that no such (x^*, t^*) exists and therefore $u(x,t) > 0$ for all $x \in \Omega$ and $t > 0$.

Finally, since u attains its minimum on $\bar{\Omega} \times [0,T]$ at $(x,t) \in \partial\Omega \times (0,T)$, Theorem 2.2 implies that the normal derivative of u is negative at such points.

Now consider the case that Robin boundary conditions hold and $\alpha_i(x) > 0$ for all $x \in \partial\Omega$. Again, if the result were false, then $u(q) < 0$ for some point q. Choose $T > 0$ such that q belongs to $\bar{\Omega} \times [0,T]$ and let u attain its minimum value on $\bar{\Omega} \times [0,T]$ at $p = (\bar{x},\bar{t})$. Since $u(p) < 0$, it follows that $\bar{t} > 0$. The argument in the previous case shows that $\bar{x}$ cannot belong to Ω. Therefore, $\bar{x} \in \partial\Omega$ and $u(x,t) > u(p)$ for all $(x,t) \in \Omega \times (0,T)$. If $\bar{t} < T$, then by Theorem 2.2, we conclude that the normal derivative of u at p is negative. But this and the fact that $u(p) < 0$ is incompatible with the boundary conditions. Even if $\bar{t} = T$, it still follows that $\frac{\partial u}{\partial \nu}(p) \leq 0$ and since $\alpha_i(\bar{x}) > 0$, the boundary conditions are violated. Therefore, $u(x,t) \geq 0$ for all $(x,t) \in \bar{\Omega} \times [0,\infty)$. Suppose now that $\phi_i(\bar{x}) > 0$ for some $\bar{x} \in \bar{\Omega}$. If $u(p) = 0$ for some $p = (x^*, t^*) \in \bar{\Omega} \times (0,\infty)$, then, by the argument above, u attains its minimum on $\Omega \times (0,T)$ for some $T > t^*$ at p. If $x^* \in \Omega$, then Theorem 2.1 implies the contradiction $\phi_i(\bar{x}) = 0$. If $x^* \in \partial\Omega$, then Theorem 2.2 implies that the normal derivative of u at p is negative and therefore the boundary conditions cannot hold at p. Either way we have a contradiction so $u(x,t) > 0$ must hold for $(x,t) \in \bar{\Omega} \times (0,\infty)$.

Finally, we consider the case that Neumann or Robin boundary conditions hold but assume only that $\alpha_i(x) \geq 0$ holds at all points of $\partial\Omega$. Let $a > 0$ satisfy $\alpha_i(x) < a$ for all $x \in \bar{\Omega}$. Suppose that $\phi_i \in X_i^+$ and $\phi_i(x) > 0$ for some $x \in \bar{\Omega}$. Let u satisfy (1.2) and let v satisfy (1.2) with α_i replaced by a. Set $w = v - u$. Then w satisfies the differential equation in (1.2) and $w(x,0) = 0$ for $x \in \bar{\Omega}$. On $\partial\Omega$, w satisfies

$$\frac{\partial w}{\partial n}(x,t) + av(x,t) - \alpha_i(x)u(x,t) = 0. \tag{2.2}$$

If $w(q) > 0$ for some $q \in \bar{\Omega} \times (0,\infty)$, let $T > 0$ be such that $q \in \bar{\Omega} \times [0,T]$ and suppose that w attains its maximum on $\bar{\Omega} \times [0,T]$ at the point $p = (\bar{x}, \bar{t})$. Obviously, $w(p) > 0$ and therefore $\bar{t} > 0$. By Theorem 2.1 applied to w, p cannot belong to $\Omega \times (0,T]$ since w vanishes on $\bar{\Omega} \times \{0\}$ and $w(p) > 0$. Thus $\bar{x} \in \partial\Omega$ and consequently $\frac{\partial w}{\partial \nu}(p) \geq 0$. But then (2.2) implies that $av(p) - \alpha_i(\bar{x})u(p) \leq 0$, contradicting that $w(p) = v(p) - u(p) > 0$ and $a > \alpha_i(\bar{x})$. Therefore, $w(x,t) = v(x,t) - u(x,t) \leq 0$ for $(x,t) \in \bar{\Omega} \times [0,\infty)$. Since $\phi_i(x) > 0$, we have from the previous paragraph that $0 < v(x,t) \leq u(x,t)$ for $(x,t) \in \bar{\Omega} \times (0,\infty)$. This and the continuity of $T_i(t)$ on X_i complete the proof. $\square$

Similar arguments show that certain order intervals $[a,b]$ are positively invariant for (1.2) in the sense that if ϕ_i takes all its values in the interval, then so does u_i.

COROLLARY 2.4. *Let $a,b \in \mathbb{R}$ satisfy $a < b$ and, in the case of Dirichlet or Robin boundary conditions, assume that $a \leq 0 \leq b$. If $\Gamma = \{\phi_i \in X_i : a \leq \phi_i(x) \leq b, x \in \bar{\Omega}\}$, then $T_i(t)\Gamma \subset \Gamma$ for $t \geq 0$. In particular, $|T_i(t)| \leq 1$ for $t \geq 0$.*

PROOF. Let $\phi_i \in \Gamma$ and let $u(x,t) = u_i(x,t) = [T_i(t)\phi_i](x)$. First consider the case of Dirichlet boundary conditions where $a \leq 0 \leq b$. Suppose for contradiction that $u(q) < a$ for some point q. Let $T > 0$ be such that $q \in \Omega \times (0,T)$ and let the minimum value of u on $\bar{\Omega} \times [0,T]$ be attained at $p = (\bar{x}, \bar{t}) \in \bar{\Omega} \times [0,T]$. Clearly, $u(p) < a$, $\bar{t} > 0$ and $\bar{x} \in \Omega$. By Theorem 2.1, $u(x,t) = u(p) < a$ for all $(x,t) \in \Omega \times (0,T)$ with $t \leq \bar{t}$. Therefore, $\phi_i(x) < a$, a contradiction. A similar contradiction results if it is assumed that $u(q) > b$ for some $q \in \bar{\Omega} \times [0,T]$.

Suppose that Robin boundary conditions hold, $a \leq 0 \leq b$ and that $\alpha_i(x) > 0$ on $\partial\Omega$. If $u(q) > b$ for some point q, choose $T > 0$ such that $q \in \bar{\Omega} \times [0,T]$ and assume that u attain its maximum on $\bar{\Omega} \times [0,T]$ at p. Obviously, $u(p) > b$. The point p cannot belong to $\Omega \times (0,T]$ since Theorem 2.1 would imply the contradiction that $u(x,0) = \phi_i(x) = u(p) > b$. Therefore, $p = (\bar{x}, \bar{t})$ with $\bar{x} \in \partial\Omega$. If $\bar{t} < T$, then Theorem 2.2 implies that the normal derivative of u at p is positive. Therefore,

$$0 = \alpha_i(\bar{x})u(p) + \frac{\partial u}{\partial \nu}(p) > \alpha_i(\bar{x})u(p) > 0$$

since $u(p) > b \geq 0$. Even if $\bar{t} = T$, the normal derivative must be nonnegative and the same contradiction holds. A similar argument shows that $u(x,t) \geq a$.

Now we drop the assumption that $\alpha_i(x) > 0$ and use the comparison, established in the proof of Corollary 2.3, that $v(x,t) \leq u(x,t)$ if v is the solution to (1.2) but with α_i replaced by a constant α where $\alpha > \alpha_i(x)$ for all $x \in \partial\Omega$. Since $v(x,t) \geq a$ by the argument in the previous paragraph, we conclude that $u(x,t) \geq a$ as well. Thus, if $a \leq 0$ and $a \leq u(x,0)$ for $x \in \bar{\Omega}$, then $a \leq u(x,t)$ for $x \in \bar{\Omega}$ and $t \geq 0$. In particular, if $0 \leq b$ and $-b \leq -u(x,0)$ for $x \in \bar{\Omega}$, then $-b \leq -u(x,t)$ for all $x \in \bar{\Omega}$ and $t \geq 0$. Therefore, $u(x,t) \leq b$ for $x \in \bar{\Omega}$ and $t \geq 0$ if $u(x,0) \leq b$ for all $x \in \bar{\Omega}$ and $b \geq 0$.

In order to establish the final assertion of the corollary, given $\phi_i \in X_i$, let $a = -|\phi_i|$ and $b = |\phi_i|$ and Γ be as above. Since Γ is positively invariant, it follows that $|T_i \phi_i| \leq |\phi_i|$. □

Theorems 2.1 and 2.2 deal with a single parabolic differential inequality while we are faced with the system (1.1) of parabolic equations. It should come as no surprise then that we need a maximum principle for systems of parabolic differential inequalities. We conclude the present section by formulating such a result. Consider the following system of weakly coupled parabolic differential inequalities

$$A_i u_i + \sum_{j=1}^{n} h_{ij} u_j \geq 0, \quad 1 \leq i \leq n. \tag{2.3}$$

where A_i is a uniformly parabolic differential operator of the form (2.1) with coefficients that are continuous on $\bar{\Omega} \times [0,T]$ for each i and $h_{ij} = h_{ij}(x,t)$ is continuous on $\bar{\Omega} \times [0,T]$ and satisfies

$$h_{ij}(x,t) \geq 0, \quad i \neq j, \quad (x,t) \in \bar{\Omega} \times [0,T].$$

If u_i, $1 \leq i \leq n$, satisfies (2.3), we let $u = (u_1, u_2, \ldots, u_n)$. The next result is crucial for establishing differential inequality and monotonicity results for systems.

THEOREM 2.5. *Suppose that each component of the vector function $u(x,t)$ satisfies the smoothness hypotheses of both Theorems 2.1 and 2.2 and that (2.3) holds for $(x,t) \in \Omega \times (0,T]$. Let $\bar{M} \in \mathbb{R}^n$ be a vector all components of which equal $M \geq 0$. If $u(x,t) \leq \bar{M}$ for $t = 0$ and $x \in \Omega$ and for $(x,t) \in Q$, then $u(x,t) \leq \bar{M}$ for $(x,t) \in \Omega \times (0,T)$. Furthermore, if $u_i = M$ at a point $(x^*, t^*) \in \Omega \times (0,T]$, then $u_i(x,t) = M$ for all (x,t) with $t \leq t^*$ and $x \in \bar{\Omega}$. If $u \leq \bar{M}$ in $\bar{\Omega} \times [0,T]$, $u(x,t) \ll \bar{M}$ for $(x,t) \in \Omega \times (0,T)$, and $u_i(x^*, t^*) = M$ for some $(x^*, t^*) \in Q$, then $\frac{\partial u_i}{\partial \nu}(x^*, t^*) > 0$.*

REMARK 2.1. Technically, Theorem 2.5 is not correct as stated since an additional hypothesis is required in the case that $M > 0$ which is not included in the statement of the theorem. The missing hypothesis is that

$$\sum_{j=1}^{n} h_{ij}(x,t) \leq 0, \quad 1 \leq i \leq n.$$

We have not included this restriction since, in all the applications to follow, it can be ignored because on replacing u_i by $\bar{u}_i = u_i e^{-ct}$ for sufficiently large $c > 0$, it is satisfied. See Protter and Weinberger(1967), Chapter 3, Theorem 15.

3. Positively Invariant Sets, Comparison and Monotonicity

In this section, we return to the consideration of (1.1) with the aim of establishing sufficient conditions for solutions to exist and remain in certain closed convex subsets of X. We also wish to make comparisons between solutions of (1.1) and solutions of various differential inequalities related to (1.1) and finally, to provide conditions for (1.1) to generate a monotone semiflow. Many of the results of the present section are analogs of similar ones for ordinary and delay differential equations established in previous chapters. The first result establishes that mild solutions exist and are, in fact, classical solutions. In addition, it summarizes many well-known facts concerning solutions of (1.1).

Let Λ be a nonempty closed convex subset of $\mathbb{R}^n$ and X_Λ be the subset of X consisting of functions ϕ which take all their values in Λ.

$$X_\Lambda = \{\phi \in X : \phi(x) \in \Lambda, \ x \in \bar{\Omega}\}.$$

In the case of Dirichlet boundary conditions, X_Λ is nonempty if and only if $0 \in \Lambda$. Sufficient conditions for X_Λ to be positively invariant with respect to the semiflow generated by (1.1) are given below. The first is the well-known Nagumo condition for the positive invariance of Λ for the parameter-dependent ordinary differential equation $u'(t) = F(x, u(t))$:

$$\lim_{h \to 0+} h^{-1} \text{dist}(\Lambda, v + hF(x,v)) = 0, \quad (x,v) \in \bar{\Omega} \times \Lambda. \tag{3.1}$$

The second condition requires that the linear semigroup $T(t)$ leaves X_Λ positively invariant:

$$T(t)X_\Lambda \subset X_\Lambda, \ t \geq 0. \tag{3.2}$$

Together, these conditions are sufficient for the positive invariance of X_Λ.

THEOREM 3.1. *Suppose that (3.1) and (3.2) hold. Then for each $\phi \in X_\Lambda$, (1.1) has a unique noncontinuable mild solution $u(t) = u(t, \phi) \in X_\Lambda$ defined on $[0, \sigma)$ where $\sigma = \sigma(\phi) \leq \infty$. Furthermore, the following properties hold:*
(a) *$u(t)$ is continuously differentiable on $(0, \sigma)$, $u(t) \in D(A)$, and $u(t)$ satisfies (1.4) on $(0, \sigma)$.*
(b) *$u(x,t) \equiv [u(t)](x)$ is a classical solution of (1.1).*
(c) *If $\sigma < \infty$, then $|u(t)| \to \infty$ as $t \to \sigma$.*
(d) *If $\sigma(\phi) = +\infty$ for all $\phi \in X_\Lambda$, then $\Phi_t(\phi) = u(t, \phi)$ is a semiflow on X_Λ.*
(e) *If B is a closed and bounded subset of X_Λ, $t_0 > 0$ and $\cup_{0 \leq t \leq t_0} \Phi_t(B)$ is bounded, then $\Phi_{t_0}(B)$ has compact closure in X_Λ.*

PROOF. The existence of a unique local mild solution of (1.5), belonging to X_Λ, is a consequence of Theorem 2.1, Chapter 8, of Martin(1976). The uniqueness follows from the smoothness hypothesis on F since it is Lipschitz in its second variable. That the solution can be continued to a maximal interval of existence $[0, \sigma)$ such that (c) holds is a consequence of Proposition 4.1 of Martin(1976). Theorem 3.1 of the same reference shows that $u(t)$ is a solution of (1.4) and (a) holds. This result requires the analyticity of the semigroup $T(t)$ as well as the hypothesis that F is Lipschitz in its second argument. The fact that $\Phi_t(\phi) = u(t, \phi)$ defines a semiflow on X_Λ when $\sigma(\phi) = +\infty$ for all $\phi \in X_\Lambda$ and that (e) holds follows from Theorem 5.2 of the above-mentioned reference. The compactness of the semigroup $T(t)$ is the key to (e) holding.

Assertion (b) involves a kind of bootstrap argument. Having shown that the unique mild solution of (1.5) is in fact a solution of (1.4), we conclude that $u(t) \in D(A)$ and $u'(t)$ exists and is continuous for $t > 0$. The latter implies that $\frac{\partial u}{\partial t}(x, t)$ is continuous for $(x, t) \in \bar{\Omega} \times (0, \sigma)$. The former means that for each $t > 0$, the components of $u(t)$ belong to $C^1(\bar{\Omega}) \cap W^{2,p}$ for all $p > N$, where $W^{2,p}$ denotes the usual space of functions $u \in L^p(\Omega)$ which have distributional derivatives up to second order which belong to $L^p(\Omega)$. See e.g. Pazy(1983) or Stewart(1980) for the above assertions. Using the smoothness assumptions on F and the fact that $u(x, t) = [u(t, \phi)](x)$ is continuously differentiable in $x \in \bar{\Omega}$, it follows that for each ϵ satisfying $0 < \epsilon < \sigma$, the function $h(x, t) \equiv F(x, u(x, t))$ is continuously differentiable for $(x, t) \in \bar{\Omega} \times [\epsilon, \sigma - \epsilon]$. Thus h is Lipschitz on this set. Now, the system

$$\frac{\partial v}{\partial t}(x, t) = D\Delta v(x, t) + h(x, t), \quad x \in \Omega, t > \epsilon$$

$$0 = \alpha(x)v(x, t) + \beta \frac{\partial v}{\partial \nu}(x, t), \quad t > \epsilon, \ x \in \partial\Omega$$

$$v(x, \epsilon) = u(x, \epsilon), \quad x \in \Omega$$

has a unique mild solution, namely $v(t) = u(t + \epsilon)$. We establish below that it has a smooth classical solution which is also a mild solution and therefore, must agree with $u(t + \epsilon)$. The proof of the existence of a classical solution of the system above is different for Dirichlet boundary conditions than for Robin conditions. Therefore, we consider them separately below.

Consider first the case of Robin boundary conditions. The existence of a classical solution v of the initial-boundary value problem above on $\bar{\Omega} \times [\epsilon, \sigma - \epsilon]$ follows from Corollary 2 to Theorem 2 in Chapter 5 of Friedman(1964) and uses the fact that $u(x, \epsilon)$ is continuously differentiable in $\bar{\Omega}$. Note also the Corollary concluding section 4, Chapter 5 of the above reference which ensures that the boundary conditions are satisfied in the stronger sense used here as opposed to the sense of (3.4), section 3, Chapter 5 of the reference. The second derivatives of v with respect to x and the first derivatives with respect to both variables x and t satisfy a local Hoelder condition in (x, t) by the interior estimates contained in Theorem 5, Chapter 3 of Friedman(1964). It is then easy to see that v is also a mild solution of the initial-boundary value problem and so must agree with $[u(t + \epsilon)](x)$. This shows that $u(x, t)$ is a classical solution of (1.1) since $\epsilon > 0$ was arbitrary.

In the case of Dirichlet boundary conditions, the existence and smoothness of a classical solution of the initial-boundary value problem is contained in Corollary

2 to Theorem 9, Chapter 3 of Friedman(1964). See also Friedman(1965). The remainder of the argument is exactly as above. Thus we have proved (b). □

The nonnegative cone $X_+ = \prod_1^n X_i^+$ in X is just X_Λ where $\Lambda = \mathbb{R}_+^n$. For many applications, it is the natural domain for (1.1) and familiar conditions on F ensure that it is positively invariant.

COROLLARY 3.2. *Let $\Lambda = \mathbb{R}_+^n$ and suppose that $F : \bar{\Omega} \times \mathbb{R}_+^n \to \mathbb{R}^n$ satisfies*

$$F_i(x,u) \geq 0 \text{ whenever } x \in \bar{\Omega}, u \in \mathbb{R}_+^n \text{ and } u_i = 0. \tag{3.3}$$

Then (3.1) and (3.2) hold for $X_\Lambda = X_+$ so the conclusions of Theorem 3.1 hold.

PROOF. We show that (3.3) implies (3.1). (3.2) follows by Corollary 2.3. Let $(x,v) \in \bar{\Omega} \times \mathbb{R}_+^n$. If $v_i > 0$, then $v_i + hF_i(x,v) > 0$ for all small $h > 0$; if $v_i = 0$, then $v_i + hF_i(x,v) = hF_i(x,v) \geq 0$ by (3.3) for all $h > 0$. Therefore, $v + hF(x,v) \in \mathbb{R}_+^n$ for all small $h > 0$. This obviously implies that (3.1) holds. □

For $a \in \mathbb{R}^n$, let $\hat{a} \in C(\bar{\Omega}, \mathbb{R}^n)$ be the constant function taking the value a. Note that $\hat{a} \notin X$ unless $a = 0$ in the case of Dirichlet boundary conditions. The usual conditions on F are sufficient for the positive invariance of the order interval $[\hat{a}, \hat{b}]$ in case of Neumann boundary conditions. In the case of Dirichlet or Robin boundary conditions, we must also assume that $a \leq 0 \leq b$.

COROLLARY 3.3. *Let $\Lambda = [a,b]$ where $a, b \in \mathbb{R}^n$ satisfy $a \ll b$. Suppose that $F : \bar{\Omega} \times [a,b] \to \mathbb{R}^n$ has the property that whenever $x \in \bar{\Omega}$ and $u \in [a,b]$ satisfies $u_i = a_i$ ($u_i = b_i$), then*

$$F_i(x,u) \geq 0 \ (F_i(x,u) \leq 0).$$

In case of Dirichlet or Robin boundary conditions, assume also that $a \leq 0 \leq b$. Then the conclusions of Theorem 3.1 hold for $X_\Lambda = [\hat{a}, \hat{b}]$.

PROOF. The proof is similar to the proof of the previous corollary. We first verify that (3.1) holds. Let $(x,v) \in \bar{\Omega} \times [a,b]$. If $a_i < v_i < b_i$, then $a_i < v_i + hF_i(x,v) < b_i$ for all small $h > 0$; if $a_i = v_i$, then $v_i + hF_i(x,v) = a_i + hF_i(x,v) \in [a_i, b_i]$ for all small $h > 0$ since $F_i(x,v) \geq 0$; a similar conclusion holds if $v_i = b_i$. Therefore, $v + hF(x,v) \in [a,b]$ for all small $h > 0$. This implies that (3.1) holds. (3.2) holds by Corollary 2.4. □

Now consider (1.1) where Λ is some nonempty, closed, convex subset of $\mathbb{R}^n$ such that the hypotheses of Theorem 3.1 hold. The set Λ is fixed through out the following. Our aim is to compare solutions of (1.1) with solutions of related differential inequalities. Let $v^+(x,t)$ and $v^-(x,t)$ be continuous on $\bar{\Omega} \times [0,\tau)$, continuously differentiable on $\bar{\Omega} \times (0,\tau)$ and twice continuously differentiable in $x \in \Omega$ for $t > 0$. Furthermore, assume that

$$v^-(x,t) \leq v^+(x,t) \text{ and } v^-(x,t), v^+(x,t) \in \Lambda, \ (x,t) \in \bar{\Omega} \times [0,\tau).$$

Let $F^\pm : \bar{\Omega} \times \Lambda \to \mathbb{R}^n$ be two functions satisfying

$$\frac{\partial F_i^\pm}{\partial u_j}(x,u) \geq 0, \quad (x,u) \in \bar{\Omega} \times \Lambda, \ i \neq j.$$

When this holds, we say that $F^\pm$ is *cooperative*. In the literature of partial differential equations it is more common to refer to this condition as the *quasimonotone* condition and it is for this reason that "quasimonotone" appears in the chapter title rather than "cooperative". Finally, assume that $v^\pm$ satisfy the differential inequalities

$$\frac{\partial v^+}{\partial t} \geq D\Delta v^+ + F^+(x, v^+), \ t > 0, x \in \Omega$$
$$\alpha v^+ + \beta \frac{\partial v^+}{\partial \nu} \geq 0, \ t > 0, x \in \partial\Omega$$
(3.4+)

and

$$\frac{\partial v^-}{\partial t} \leq D\Delta v^- + F^-(x, v^-), \ t > 0, x \in \Omega$$
$$\alpha v^- + \beta \frac{\partial v^-}{\partial \nu} \leq 0, \ t > 0, x \in \partial\Omega$$
(3.4−)

The function v^+ is called a *super-solution* and v^- is called a *sub-solution* in the literature of partial differential equations. The next result is a fundamental comparison technique.

THEOREM 3.4. *Suppose that $v^\pm$ satisfy (3.4±) and $F^\pm$ are cooperative. Suppose further that*

$$F^-(x, u) \leq F(x, u) \leq F^+(x, u), \quad (x, u) \in \bar{\Omega} \times \Lambda$$

and $\phi \in X_\Lambda$ satisfies

$$v^-(x, 0) \leq \phi(x) \leq v^+(x, 0), \quad x \in \bar{\Omega}.$$

Then the unique solution of (1.1) exists on $[0, \sigma)$ where $\sigma > \tau$ and

$$v^-(x, t) \leq u(x, t) \leq v^+(x, t), \quad (x, t) \in \bar{\Omega} \times [0, \tau).$$

PROOF. We give a complete proof for Dirichlet boundary conditions and for Robin boundary conditions when $\alpha(x) > 0$ for all $x \in \partial\Omega$. We refer the reader to Fife(1979), Fife and Tang(1981) or Leung(1989) for a proof in the remaining case. See also Martin and Smith(1990) for a proof in case time delays appear in the systems.

Let $u(x, t)$ be the solution of (1.1) guaranteed by Theorem 3.1 and let $w(x, t) = u(x, t) - v^+(x, t)$. Then $w(x, t)$ satisfies the differential inequality

$$0 \leq D\Delta w - \frac{\partial w}{\partial t} + F(x, u) - F^+(x, v^+).$$

Using the inequality

$$F(x, u) - F(x, v^+) \leq F^+(x, u) - F^+(x, v^+) = \int_0^1 \frac{\partial F^+}{\partial u}(x, su + (1-s)v^+) ds \ w,$$

we see that w satisfies a differential inequality of the form (2.3).

Consider first the case of Dirichlet boundary conditions. In that case, w satisfies $w(x, t) \leq 0$ for $t = 0$ and $x \in \Omega$ and for $x \in \partial\Omega$ and $t > 0$. By Theorem 2.5, we conclude that $w(x, t) \leq 0$ for $x \in \Omega$ and $t > 0$. Equivalently, $u(x, t) \leq v^+(x, t)$ as asserted.

Now assume that the Robin boundary conditions hold where $\alpha(x) > 0$ for all $x \in \partial\Omega$. Now, $\dot{w} \leq 0$ holds for $t = 0$ and
$$\alpha w + \frac{\partial w}{\partial \nu} \leq 0, \quad x \in \partial\Omega, t > 0.$$
Suppose that $w_i(\bar{x}, \bar{t}) > 0$ for some $\bar{x} \in \bar{\Omega}$ and $\bar{t} > 0$. Let $M = \max\{w_j(x,t) : (x,t) \in \bar{\Omega} \times [0, \bar{t}], 1 \leq j \leq n\}$ and suppose that $M = w_l(x^*, t^*)$ for some l and $(x^*, t^*) \in \bar{\Omega} \times [0, \bar{t}]$. Since $M > 0$, by Theorem 2.5, if $x^* \in \Omega$, then $w_l(x,t) = M > 0$ for all $t \leq \bar{t}$ which contradicts the fact that $w(x,0) \leq 0$. Therefore, the value M cannot be attained by any component of w at any point $(x,t) \in \Omega \times [0, \bar{t}]$. Consequently, $x^* \in \partial\Omega$ and we may conclude that $\frac{\partial w_l}{\partial \nu}(x^*, t^*) \geq 0$. This together with the fact that $w_l(x^*, t^*) > 0$, leads to a contradiction of the boundary condition
$$\alpha_l(x^*)w_l(x^*, t^*) + \frac{\partial w_l}{\partial \nu}(x^*, t^*) \leq 0.$$
Therefore, $w(x,t) \leq 0$ and so $u(x,t) \leq v^+(x,t)$ holds for $(x,t) \in \bar{\Omega} \times [0, \tau)$. Similar arguments establish the inequality $v^-(x,t) \leq u(x,t)$. $\square$

If, in Theorem 3.4, we assume only the existence of v^+ (v^-) satisfying (3.4+) ((3.4-)) and only the inequalities involving v^+ (v^-) are retained in the statement of the theorem, then the inequality $u(x,t) \leq v^+(x,t)$ ($v^-(x,t) \leq u(x,t)$) holds but only on the common interval of existence of the two functions.

Theorem 3.4 can be used to show that (1.1) generates a monotone semiflow if F is cooperative. The cone X_+ generates a partial order on X in the usual way. If $\phi, \psi \in X$, then $\phi \leq \psi$ if and only if $\phi(x) \leq \psi(x)$ for all $x \in \bar{\Omega}$. As usual, $\phi < \psi$ if $\phi \leq \psi$ and $\phi \neq \psi$. The choice of $\Lambda = \mathbb{R}^n_+$ below is merely for convenience.

COROLLARY 3.5. *Suppose that $F : \bar{\Omega} \times \mathbb{R}^n_+ \to \mathbb{R}^n$ satisfies (3.3) and is cooperative. If $\phi, \psi \in X_+$ satisfy $\phi \leq \psi$ and $u(t, \phi), u(t, \psi)$ are the corresponding solutions of (1.1) satisfying $u(0, \phi) = \phi, u(0, \psi) = \psi$, defined on $[0, \sigma(\phi))$ and $[0, \sigma(\psi))$ respectively, then $\sigma(\phi) \geq \sigma(\psi)$ and*
$$u(t, \phi) \leq u(t, \psi), \quad 0 \leq t < \sigma(\psi).$$
If $\phi < \psi$, then
$$u(t, \phi) < u(t, \psi), \quad 0 < t < \sigma(\psi).$$

PROOF. Take $F^\pm = F$ and $v^- = 0$. By (3.3), v^- satisfies (3.4-). Corollary 3.2 implies that $u(t, \psi) \in X_+$ is defined on a maximal interval $[0, \sigma(\psi))$. If $v^+(x,t) = [u(t, \psi)](x)$, then by Theorem 3.4, there is a unique noncontinuable solution $u(t, \phi)$ defined on $[0, \sigma(\phi))$, where $\sigma(\phi) > \sigma(\psi)$, such that $0 \leq u(t, \phi) \leq u(t, \psi)$ for $0 \leq t < \sigma(\psi)$. This proves the first assertion.

Now suppose that $\phi < \psi$ and, as in the proof of Theorem 3.4, let $w(\bullet, t) = u(t, \phi) - u(t, \psi)$. Then $w(x,t)$ satisfies $w(x,t) \leq 0$ by the preceding paragraph and w satisfies a differential inequality of the form (2.3) by an argument similar to that used in Theorem 3.4. If $w(x, t_0) = 0$ for all $x \in \bar{\Omega}$ for some $t_0 > 0$, then, by Theorem 2.5, $w(x,t) = 0$ for all $(x,t) \in \Omega \times [0, t_0)$. In this case it follows that $\phi = \psi$. Since $\phi < \psi$, this contradiction proves the final assertion. $\square$

The last assertion of Corollary 3.5 implies that (1.1) generates a strictly monotone semiflow.

The existence of monotone solutions for (1.1), or equivalently, of sub-equilibria and super-equilibria (see section 5 of Chapter 2), can be shown using Theorem 3.4. The next result should be compared to Proposition 3.2.1 for ordinary differential equations and Corollary 5.2.2 for delay differential equations. See also Lemma 2.5.1.

COROLLARY 3.6. *Suppose that $F : \bar{\Omega} \times \mathbb{R}_+^n \to \mathbb{R}^n$ satisfies (3.3) and is cooperative. Let $\phi^- : \bar{\Omega} \to \mathbb{R}_+^n$ satisfy $\phi^- \in C^1(\bar{\Omega}, \mathbb{R}^n) \cap C^2(\Omega, \mathbb{R}^n)$ and*

$$D\Delta\phi^-(x) + F(x, \phi^-(x)) \geq 0, \quad x \in \Omega$$
$$\alpha(x)\phi^-(x) + \beta\frac{\partial\phi^-}{\partial\nu}(x) \leq 0, \quad x \in \partial\Omega. \tag{3.5}$$

Then for $0 \leq t_1 \leq t_2 < \sigma(\phi^-)$ and $x \in \bar{\Omega}$, the solution $u(x,t) = [u(t, \phi^-)](x)$ of (1.1) satisfies

$$\phi^-(x) \leq u(x, t_1) \leq u(x, t_2), \quad x \in \bar{\Omega}. \tag{3.6}$$

Furthermore, if $u(t, \phi^-)$ is bounded from above in X, that is, if there exists $M \in X_+$ such that $u(t, \phi^-) \leq M$ for $t \in [0, \sigma(\phi^-))$, then $\sigma(\phi^-) = +\infty$ and there exists $v \in C^1(\bar{\Omega}, \mathbb{R}^n) \cap C^2(\Omega, \mathbb{R}^n)$ satisfying

$$0 = D\Delta v(x) + F(x, v(x)), \quad x \in \Omega$$
$$0 = \alpha(x)v(x) + \beta\frac{\partial v}{\partial\nu}(x), \quad x \in \partial\Omega, \tag{3.7}$$

such that

$$u(t, \phi^-) \to v, \quad t \to \infty.$$

Similarly, if $\phi^+ : \bar{\Omega} \to \mathbb{R}_+^n$ satisfies the same smoothness conditions as ϕ^- and if (3.5) holds for ϕ^+ with inequalities reversed, then the solution $u(x, t) = [u(t, \phi^+)](x)$ of (1.1) satisfies (3.6) with ϕ^+ in place of ϕ^- and with inequalities reversed. Furthermore, since $0 \leq u(t, \phi^+)$, there exists a solution v of (3.7) such that $u(t, \phi^+) \to v$ in X as $t \to \infty$.

PROOF. We apply Theorem 3.4 with $v^- = \phi^-$ $F^- = F$ and $\phi = \phi^-$. Clearly, (3.5) implies that (3.4−) is satisfied. By Theorem 3.4, $u(t, \phi^-) \geq \phi^-$ on the maximal interval of existence of $u(t, \phi^-)$. Corollary 3.5 and the semigroup property then imply that $u(t + s, \phi^-) \geq u(s, \phi^-)$ holds for all $t, s \geq 0$ such that $t + s < \sigma(\phi^-)$. This establishes the monotonicity of the solution $u(t, \phi^-)$.

If $u(t, \phi^-)$ is bounded from above, then $\overline{O(\phi^-)}$ is bounded, where $O(\phi^-) = \{u(t, \phi^-) : t \geq 0\}$, and so by Theorem 3.1 (e), the orbit $O(\phi^-)$ of ϕ^- has compact closure in X_+. The monotonicity of the solution then implies that it must converge to equilibrium as $t \to \infty$ (see proof of Lemma 2.5.1). □

4. The Strong Order Preserving Property

Establishing the strong order preserving property is the goal of this section and it breaks naturally into two cases. In the case of Robin or Neumann boundary conditions, the cone X_+ has nonempty interior in X consisting of those functions ϕ satisfying $\phi(x) \gg 0$ for all $x \in \bar{\Omega}$. In this case, it is relatively easy to establish that Φ is strongly monotone if F is cooperative and irreducible (see below) by using the maximum principle, Theorem 2.5. However, in the case of Dirichlet boundary conditions, the cone X_+ has empty interior in X because $\phi(x) = 0$ for all $x \in \partial\Omega$ for each $\phi \in X_+$. In this case, we need to make use of the fact that Φ_t maps X

continuously into $C^1(\bar{\Omega}, \mathbb{R}^n)$ and use Remark 1.1 of Chapter 1 in order to prove that Φ has the strong order preserving property. In this section we assume that $\Lambda = \mathbb{R}_+^n$ merely for specificity. We also assume that $\sigma(\phi) = +\infty$ for every $\phi \in X_+$ so that Φ defines a semiflow by Theorem 3.1.

We say that $F : \bar{\Omega} \times \mathbb{R}_+^n \to \mathbb{R}^n$ is *cooperative and irreducible* if it is cooperative and there exists $\bar{x} \in \Omega$ such that the Jacobian matrix $\frac{\partial F}{\partial u}(\bar{x}, u)$ is an irreducible matrix for every $u \geq 0$. Observe that in the scalar case, $n = 1$, the hypothesis that F is cooperative and irreducible is automatically satisfied. The next result establishes that, in case of Robin or Neumann boundary conditions, Φ is strongly monotone.

THEOREM 4.1. *Let F be cooperative and irreducible and suppose that (3.3) holds. Let $\phi, \psi \in X_+$ satisfy $\phi < \psi$ and let $u(x, t, \phi)$ and $u(x, t, \psi)$ be the solutions of (1.1) satisfying $u(\bullet, 0, \phi) = \phi$ and $u(\bullet, 0, \psi) = \psi$. Then*

$$u(x, t, \phi) \ll u(x, t, \psi) \tag{4.1}$$

holds for all $x \in \Omega$ and $t > 0$ in the case of Dirichlet boundary conditions and for all $x \in \bar{\Omega}$ and $t > 0$ for Robin or Neumann boundary conditions. Moreover, in the case of Dirichlet boundary conditions we have

$$\frac{\partial u}{\partial \nu}(x, t, \psi) \ll \frac{\partial u}{\partial \nu}(x, t, \phi) \tag{4.2}$$

for all $x \in \partial\Omega$ and $t > 0$.

PROOF. Let $w(x, t) = u(x, t, \phi) - u(x, t, \psi)$. By Corollary 3.5, $w(x, t) \leq 0$ for all $x \in \bar{\Omega}$ and $t \geq 0$. Furthermore, as in the proof of Theorem 3.4, w satisfies a differential inequality (actually, an equality) of the form (2.3) where

$$h_{ij}(x, t) = \int_0^1 \frac{\partial F_i}{\partial u_j}(x, su(x, t, \phi) + (1-s)u(x, t, \psi))ds \geq 0, \quad i \neq j.$$

If $w_j(x^*, t^*) = 0$ for some $x^* \in \Omega$ and $t^* > 0$, then $w_j(x, t) = 0$ for all $x \in \bar{\Omega}$ and $0 \leq t \leq t^*$ by Theorem 2.5. Let $I = \{i : w_i(x, t) = 0$ for some $(x, t) \in \Omega \times (0, \infty)\}$ and suppose that I is nonempty. Since $\phi < \psi$ it follows that there exists j such that $w_j(x, 0) < 0$ for some $x \in \Omega$ and consequently $j \notin I$. Let J be the complement of I so that $w_j(x, t) < 0$ holds for each $j \in J$, $x \in \Omega$ and $t > 0$. It follows that there exists $t_0 > 0$ such that $w_i(x, t) = 0$ for all $i \in I$, $x \in \Omega$ and $0 \leq t \leq t_0$. Then for $i \in I$ we have from (2.3) that

$$0 = A_i w_i + \sum_{j \in J} h_{ij} w_j = \sum_{j \in J} h_{ij} w_j$$

for each $x \in \Omega$ and $0 < t \leq t_0$. Since F is cooperative, $h_{ij} \geq 0$ for $i \neq j$, and since it is also irreducible, there exists $i_0 \in I$ and $j_0 \in J$ such that $h_{i_0 j_0}(\bar{x}, t) > 0$. But $w_{j_0}(\bar{x}, t) < 0$ and this leads to the contradiction $0 = \sum_{j \in J} h_{i_0 j}(\bar{x}, t) w_j(\bar{x}, t) < 0$, and proves that I is empty. Thus (4.1) holds for all $x \in \Omega$ and $t > 0$.

If Dirichlet boundary conditions hold, then $w(x, t) = 0$ for all $x \in \partial\Omega$ and $t > 0$ and consequently $\frac{\partial w}{\partial \nu}(x, t) \gg 0$ holds for all such (x, t) by Theorem 2.5. This proves (4.2).

If Robin or Neumann boundary conditions hold and $w_i(x^*, t^*) = 0$ for some i, $x^* \in \partial\Omega$ and $t^* > 0$, then the normal derivative of w_i at that point is positive by Theorem 2.5. But this contradicts the boundary condition $\frac{\partial w_i}{\partial \nu}(x^*, t^*) = 0$ which holds at (x^*, t^*) as a consequence of $w_i(x^*, t^*) = 0$. Therefore, (4.1) holds for all $x \in \bar{\Omega}$ and $t > 0$. □

In case of Dirichlet boundary conditions, $X = \prod_1^n C_0(\bar{\Omega})$, and although Φ is not strongly monotone on X_+, it is strongly order preserving as our next result shows.

COROLLARY 4.2. *Let the hypotheses of Theorem 4.1 hold and suppose that Dirichlet boundary conditions hold. Then Φ is strongly order preserving.*

PROOF. The proof is an application of Remark 1.1 of Chapter 1. Let $C^1(\bar{\Omega})$ be the Banach space of continuously differentiable functions on $\bar{\Omega}$ with the norm $|\phi|_1 = |\phi| + \sum_i |\nabla \phi_i|$. Let $C_0^1(\bar{\Omega})$ be the closed subspace of $C^1(\bar{\Omega})$ consisting of functions vanishing on $\partial \Omega$. Finally, let $Z = \prod_1^n C_0^1(\bar{\Omega})$ and $Z_+ = Z \cap X_+$. Obviously, the inclusion map $i : Z \to X$ is a continuous linear map. Moreover, the cone Z_+ has nonempty interior in Z given by

$$\mathrm{Int} Z_+ = \{u \in Z_+ : 0 \ll u(x), x \in \Omega, \frac{\partial u}{\partial \nu}(x) \ll 0, x \in \partial \Omega\}.$$

We write $z_1 \ll_Z z_2$ in case $z_i \in Z$ and $z_2 - z_1 \in \mathrm{Int} Z_+$. Now, as we show below, Φ_t maps X continuously into Z for each $t > 0$. Thus we may interpret Theorem 4.1 as implying that if $\phi, \psi \in X_+$ satisfy $\phi < \psi$ then $\Phi_t(\phi) \ll_Z \Phi_t(\psi)$ in Z. These facts and Remark 1.1 of Chapter 1 complete the proof.

It remains to establish that Φ_t is a continuous map from X to Z for each $t > 0$. First, we note that $T(t)\phi \in D(A) \subset Z$ for each $t > 0$ and $\phi \in X$ and if $\tau > 0$, then there exists $c = c(\tau) > 0$ such that

$$|T(t)\phi|_1 \leq c t^{-1/2} |\phi|, \quad 0 < t \leq \tau. \tag{4.3}$$

The estimate (4.3) can be found in Mora(1983) (see (2.26)). This estimate can be used to show that the integral appearing in (1.5) exists as an integral with respect to the topology of Z. Indeed, it is a straightforward exercise to show that the map $s \to T(t-s)f(u(s))$ is continuous from $[0, t)$ into Z by virtue of the continuity of f on X and (4.3). Furthermore, the possible singularity of this map at $s = t$ is sufficiently mild, according to (4.3), that $\int_0^t |T(t-s)f(u(s))|_1 ds \leq Mct^{1/2}$, where $M = \sup_{0 \leq s \leq t} |f(u(s))|$. From this and (1.5), it follows that $u(t) \in Z$ for $t > 0$. Now fix $t > 0$ and let $\phi_n \to \phi$ in X_+. Using (1.5), the fact that f is Lipschitz on bounded subsets of X_+ and (4.3), we have

$$|u(t,\phi) - u(t,\phi_n)|_1 \leq c t^{-1/2} |\phi - \phi_n| + L c t^{1/2} \sup_{0 \leq s \leq t} |u(s,\phi) - u(s,\phi_n)|$$

where L is a suitable Lipschitz constant for f. Since $u(s, \phi_n) \to u(s, \phi)$ in X uniformly on $0 \leq s \leq t$ by a Gronwall argument, this estimate proves the continuity assertion. □

5. Generic Convergence for Cooperative and Irreducible Systems

We now have the tools to apply the theory of Chapters 1 and 2 to (1.1). Our aim is to show that the generic solution of (1.1) converges to an equilibrium if F is cooperative and irreducible and some additional hypotheses hold. As in the previous section, it is convenient to separate the case that Robin or Neumann

5. GENERIC CONVERGENCE FOR COOPERATIVE AND IRREDUCIBLE SYSTEMS

boundary conditions hold from the case that Dirichlet boundary conditions hold. We begin with the former case and therefore set $X = \prod_1^n C(\bar{\Omega})$.

For definiteness, we assume that $\Lambda = \mathbb{R}_+^n$ so that $X_\Lambda = X_+$. The usual compactness assumption is also required:

(T) For each $\phi \in X_+$, the solution $u(t, \phi)$ of (1.1) is defined and bounded for $t \geq 0$. Moreover, for each compact subset K of X_+, there exists a closed and bounded subset $B = B(K)$ of X_+ such that for each $\phi \in K$, $u(t, \phi) \in B$ for all large t.

One of the main result of this chapter is the following.

THEOREM 5.1. *Let $F : \bar{\Omega} \times \mathbb{R}_+^n \to \mathbb{R}^n$ be cooperative and irreducible and satisfy (3.3). Assume that the boundary conditions for (1.1) are Robin or Neumann type and that (T) holds. Then the set of convergent points in X_+ contains an open and dense subset of X_+.*

PROOF. To apply Theorem 2.4.7, we must verify the hypotheses (M),(D),(S) and (C). (M) is an immediate consequence of Theorem 4.1. We show that (C) is a consequence of (T). Since $O(\phi)$ is bounded for each $\phi \in X_+$ by (T), its closure is compact in X_+ by Theorem 3.1 (e), where we take $B = \overline{O(\phi)}$. Thus, $\omega(\phi)$ exists and has the usual properties. Now if ϕ_n approximates $\phi_0 \in X_+$ from above or from below, then $K = \cup_{n \geq 0} \{\phi_n\}$ is compact so by (T), there is a closed and bounded subset B such that $u(t, \phi_n) \in B$ for all large t. It follows that $\omega(\phi_n) \subset B$ and therefore, $\cup_{n \geq 0} \omega(\phi_n) \subset B$ and since $\cup_{n \geq 0} \omega(\phi_n)$ is invariant, it has compact closure by Theorem 3.1 (e).

In order to verify (D), it suffices to show that $\Phi_1 : X_+ \to X_+$ is continuously differentiable. The differentiability is a consequence Theorem 4.1 of Mora(1983) and our smoothness hypotheses. In fact, Mora shows that Φ_t is a C^1 semiflow on X_+ in the sense that the map Φ_t is C^1 for each $t > 0$. Furthermore, if v is an equilibrium for (1.1), that is, a solution of (1.6), then $z(x,t) = [\Phi_t'(v)\phi](x)$ satisfies the variational equation

$$\frac{\partial z}{\partial t} = D\Delta z + \frac{\partial F}{\partial u}(x, v(x))z, \quad x \in \Omega, t > 0,$$
$$0 = \alpha(x)z(x,t) + \beta \frac{\partial z}{\partial \nu}(x,t), \quad x \in \partial\Omega, t > 0, \qquad (5.1)$$
$$z(x,0) = \phi(x), \quad x \in \Omega.$$

Indeed, the linear system (5.1) generates a compact semigroup $S(t)$ of operators on X since the abstract ordinary differential equation in X representing (5.1) takes the form of (1.4) where f is a bounded linear operator. See Pazy(1983), Proposition 3.1.4. Thus, $\Phi_t'(v) = S(t)$. Corollary 3.2 can be applied to (5.1) to show that $S(t)$ is a positive semigroup, that is, $S(t)X_+ \subset X_+$ holds for $t \geq 0$ and, since F is cooperative and irreducible, the proof of Theorem 4.1 shows that $S(t)$ satisfies $0 \ll S(t)\phi$ if $0 < \phi$ and $t > 0$. In other words, $S(t)$ is strongly positive. This shows that (S) and (D) hold and completes the proof. □

Although the statement of the analog of Theorem 5.1 for Dirichlet boundary conditions is almost identical to Theorem 5.1, its proof is somewhat different. In this case $X = \prod_1^n C_0(\bar{\Omega})$.

THEOREM 5.2. *Let* $F : \bar{\Omega} \times \mathbb{R}_+^n \to \mathbb{R}^n$ *be cooperative and irreducible and satisfy* (3.3). *Assume Dirichlet boundary conditions for* (1.1) *and that* (T) *holds. Then the set of convergent points in* X_+ *contains an open and dense subset of* X_+.

PROOF. To apply Remark 4.1 to Theorem 2.4.7, we must verify the conditions (C), (I),(J),(M),(D),(S). Condition (C) follows from (T) as in the proof of Theorem 5.1. (M) follows from Corollary 4.2 and (J) was verified in the proof of Corollary 4.2. Conditions (I) and (D) will hold if the restriction of Φ to the space $Z = \prod_1^n C_0^1(\bar{\Omega})$, introduced in the proof of Corollary 4.2, is a C^1 semiflow. The latter follows from our smoothness assumptions and Theorem 4.1 of Mora(1983).

It remains to verify that (S) holds. If v is an equilibrium of (1.1), then the derivative $\Phi_t'(v)$ of the map $\Phi_t : Z \to Z$ at v is the strongly continuous semigroup $S_Z(t)$ on Z obtained as the restriction to Z of the semigroup $S(t)$ on X generated by the linear system

$$\frac{\partial z}{\partial t} = D\Delta z + \frac{\partial F}{\partial u}(x, v(x))z, \quad x \in \Omega, t > 0,$$
$$0 = z(x,t), \quad x \in \partial\Omega, t > 0, \qquad (5.2)$$
$$z(x,0) = \phi(x), \quad x \in \Omega.$$

$S(t)$ is a positive semigroup by Corollary 3.2 and therefore, so is $S_Z(t)$ a positive semigroup on $Z_+ = X_+ \cap Z$. In fact, $S_Z(t)$ is strongly positive for $t > 0$ by the proof of Corollary 4.2. It is also compact by Theorem 2.4 and (2.39) of Mora(1983) and thus, (S) holds. □

The proofs of Theorem 5.2 and Theorem 4.1 show that with Dirichlet boundary conditions (1.1) generates a C^1, strongly monotone semiflow on the space $Z = \prod_1^n C_0^1(\bar{\Omega})$. Consequently, Theorem 2.4.7 could be applied to this semiflow on the smaller space Z to conclude that Z_+ contains an open and dense subset of convergent points, a slightly weaker result than Theorem 5.2. In order to see that this result is weaker, let O be the open and dense subset of Z_+ consisting of convergent points which an application of Theorem 2.4.7 would yield and let $U = \cup_{t>0}\Phi_t^{-1}(O) \subset X_+$. Since the semiflow Φ_t, generated by (1.1) on X, maps X continuously into Z for $t > 0$, U is an open subset of convergent points in X_+. As O is dense in Z_+, in the topology of Z, and Z is dense in X, in the topology of X, it follows that O is dense (but not open) in the topology of X_+. Therefore, we can conclude from this approach that the set of convergent points is dense in X_+ and that it contains an open set, namely U, but we apparently cannot conclude from these arguments alone that the set of convergent points in X_+ contains a subset that is both open and dense. The existence of just such a set is the conclusion of Theorem 5.2. The importance of the existence of an open and dense subset of convergent points in X_+, call it C, is that not only can one approximate any point of X_+ by a sequence in C but also that the property of belonging to C is a robust one (in X) in the sense that sufficiently small perturbations of points of C are again points of C.

6. Stability of Equilibria

As the typical solution of a cooperative and irreducible reaction-diffusion system (1.1) converges to equilibrium, it becomes especially important to determine

6. STABILITY OF EQUILIBRIA

the stability properties of the equilibria. This is the central topic of the present section. We will show that when F is cooperative and irreducible, then the stability of an equilibrium is determined by a real eigenvalue of an associated eigenvalue problem and that the corresponding eigenvector has certain positivity properties which depend on the boundary conditions.

Let $v(x)$ be an equilibrium solution of (1.1) and for notational convenience let

$$M(x) = \frac{\partial F}{\partial u}(x, v(x)).$$

We assume that $M(x) = (m_{ij}(x))$ satisfies

$$m_{ij}(x) \geq 0, \quad i \neq j, x \in \bar{\Omega},$$

and that $M(x_0)$ is irreducible for some $x_0 \in \Omega$, as would automatically be the case if F were cooperative and irreducible. The variational equation about $v(x)$ is given by the linear system

$$\begin{aligned}
\frac{\partial z}{\partial t} &= D\Delta z + M(x)z, \quad x \in \Omega, t > 0, \\
0 &= \alpha(x)z(x,t) + \beta\frac{\partial z}{\partial \nu}(x,t), \quad x \in \partial\Omega, t > 0, \\
z(x,0) &= \phi(x), \quad x \in \Omega.
\end{aligned} \tag{6.1}$$

It generates a compact, positive semigroup on X as noted in the proofs of the preceding section. The stability properties of the semigroup are determined by the spectrum of its generator and therefore we are lead to the eigenvalue problem

$$\begin{aligned}
\lambda w &= D\Delta w + M(x)w, \quad x \in \Omega, t > 0, \\
0 &= \alpha(x)w(x,t) + \beta\frac{\partial w}{\partial \nu}(x,t), \quad x \in \partial\Omega, t > 0,
\end{aligned} \tag{6.2}$$

Our first result says that there is a real eigenvalue of (6.2) which is larger than the real part of all other eigenvalues of (6.2). We call it the *principal eigenvalue* of (6.2). The associated eigenvector is positive and is called the *principal eigenvector* of (6.2).

THEOREM 6.1. *There exists a real eigenvalue λ_0 of (6.2) and a corresponding eigenvector $w_0(x)$ satisfying*

$$w_0(x) \gg 0$$

for all $x \in \bar{\Omega}$ in the case of Robin or Neumann boundary conditions and for all $x \in \Omega$ in case of Dirichlet boundary conditions. In the latter case, we also have

$$\frac{\partial w_0}{\partial \nu}(x) \ll 0, \quad x \in \partial\Omega.$$

If λ is any other eigenvalue of (6.2), then the real part of λ, $\Re(\lambda)$, satisfies

$$\Re(\lambda) < \lambda_0.$$

Moreover, λ_0 is an eigenvalue of algebraic multiplicity one and any other nonnegative eigenvector of (6.2) is a positive multiple of w_0.

PROOF. We consider first the case of Robin or Neumann boundary conditions. As noted in the proof of Theorem 5.1, the linear system (6.1) generates a compact, strongly positive semigroup $S(t)$ on X_+ where $X = \prod_1^n C(\bar{\Omega})$. If L denotes the generator of this semigroup, then there exists $\mu_0 \in \mathbb{R}$ such that

$$(\mu I - L)^{-1}\phi = \int_0^\infty e^{-\mu t} S(t)\phi\, dt$$

for all $\phi \in X$ and for all $\mu > \mu_0$ (see Pazy(1983)). Fix such a μ. From this formula, it follows readily that $B = (\mu I - L)^{-1}$ is compact and strongly positive. The compactness of B follows from Pazy(1983), Theorem 3.3 of Chapter 2. The positivity of B follows immediately from the formula above and the closedness of the cone. (See also Clement et al(1987), Proposition 7.1.) That it is also strongly positive follows from the fact that any $x \in \bar{\Omega}$ can be viewed as a continuous linear functional on X by $x(\phi) = \phi(x)$. Consequently, if $\phi \in X_+$ and $\phi \neq 0$, then

$$x\left(\int_0^\infty e^{-\mu t} S(t)\phi\, dt\right) = \int_0^\infty e^{-\mu t}[S(t)\phi](x)\, dt \gg 0$$

where the last integral is an integral in $\mathbb{R}^n$ and it is positive since $[S(t)\phi](x) \gg 0$ for $t > 0$. By the Krein-Rutman Theorem (Theorem 2.4.1), the spectral radius $r = \rho(B)$ is a positive eigenvalue of B and the generalized eigenspace of r is spanned by a positive eigenvector $w_0 \gg 0$. The eigenvalues λ of L are related to the eigenvalues η of B by $\lambda = \mu - \eta^{-1}$. Let $\lambda_0 = \mu - r^{-1}$. Since $Lw_0 = \lambda_0 w_0$, it follows that $S(t)w_0 = e^{\lambda_0 t} w_0$. But the Krein-Rutman Theorem applies to $S(t)$ as well and therefore its spectral radius is $e^{\lambda_0 t}$ and the corresponding eigenvector w_0 is its unique eigenvector in X_+ up to scalar multiple. If $\lambda \neq \lambda_0$ is an eigenvalue of L with corresponding eigenvector w, then $e^{\lambda t}$ is an eigenvalue of $S(t)$ with corresponding eigenvector w, by the spectral theorem. In case $e^{\lambda t} = e^{\lambda_0 t}$, then w must be a scalar multiple of w_0 since the latter spans the eigenspace by the Krein-Rutman theorem. But then obviously $\lambda = \lambda_0$, in contradiction to our hypothesis. Thus, $e^{\lambda t} \neq e^{\lambda_0 t}$ and consequently, $|e^{\lambda t}| < e^{\lambda_0 t}$, again by the Krein-Rutman theorem. It follows that $\Re(\lambda) < \lambda_0$.

In the case of Dirichlet boundary conditions, a similar analysis applies although it is convenient to work in the space $Z = \prod_1^n C_0^1(\bar{\Omega})$, used in the proof of Theorem 5.2. $S_Z(t)$ is a compact, strongly positive semigroup on Z. The proof now continues as above after noting that the spectrum of the generators of $S_Z(t)$ and $S(t)$ are the same. The assertion that B is strongly positive on Z_+ follows as in the previous case but also makes use of the fact that evaluating the normal derivative at a point of $\partial\Omega$ is a continuous linear operator on Z. Therefore, if $z = B\phi$ where $\phi \in Z_+$ satisfies $\phi > 0$ and $x \in \partial\Omega$, then $\frac{\partial z}{\partial \nu}(x) \ll 0$ by consideration of the integral as in the previous case. $\square$

The stability of the equilibrium solution v of (1.1), in the case that F is cooperative and irreducible, is determined by the principal eigenvalue λ_0.

THEOREM 6.2. *Let $v(x)$ be an equilibrium solution of (1.1) where F is cooperative and irreducible. Let λ_0 denote the principal eigenvalue of (6.2). Then v is asymptotically stable if $\lambda_0 < 0$ and it is unstable if $\lambda_0 > 0$.*

PROOF. The Theorem is an immediate consequence of Theorem 6.1 and Theorem 4.2 of Mora(1983).

REMARK 6.1. If F is independent of x and cooperative and irreducible and if Neumann boundary conditions hold, then any solution v of $F(v) = 0$ gives rise to a constant equilibrium of (1.1). In this case, the stability modulus of the derivative of F at v, $s = s(F'(v))$, is the principal eigenvalue, λ_0. The stability properties of the constant equilibrium v of (1.1) are the same as the stability properties of v as an equilibrium of the ordinary differential equation $z' = F(z)$. In order to see this, let $w \gg 0$ be the eigenvector corresponding to s, $F'(v)w = sw$. Then $w_0(x) \equiv w$ satisfies (6.2) with $\lambda = s$ and since there is only one positive eigenfunction for (6.2), it follows that w_0 is the principal eigenfunction and $\lambda_0 = s$.

It can also happen that there exist nonconstant equilibria of (1.1). However, if Ω is convex, then any nonconstant equilibrium is unstable, in fact $\lambda_0 > 0$, according to a result of Kishimoto and Weinberger(1985).

REMARK 6.2. Let $F : \bar{\Omega} \times \mathbb{R}^n_+ \to \mathbb{R}^n$ be cooperative and irreducible and satisfy (3.3), let v be an equilibrium solution of (1.1) and suppose that the principal eigenvalue λ_0 of (6.2) corresponding to v is positive. In that case, v is unstable and this instability can be made quite explicit if we apply Corollary 3.6 with $\phi^- = v(x) + \epsilon w_0(x)$, where w_0 is the principal eigenvector, or with $\phi^+ = v(x) - \epsilon w_0(x)$, provided $\epsilon > 0$ is sufficiently small. Using the positivity of w_0, we have

$$D\Delta\phi^-(x) + F(x, \phi^-(x)) = D\Delta v + F(x, v(x)) + \epsilon(D\Delta w_0 + M(x)w_0) + o(\epsilon)w_0$$
$$= \epsilon\lambda_0 w_0(x) + o(\epsilon)w_0(x)$$
$$= \epsilon(\lambda_0 + O(\epsilon))w_0(x)$$
$$\geq 0, \quad x \in \Omega,$$

where $O(\epsilon)$ and $o(\epsilon)$ represent diagonal matrices with diagonal elements having the indicated order in ϵ. Furthermore, ϕ^- satisfies the boundary conditions. Here, we have used the smoothness of F and (1.3), if Dirichlet boundary conditions hold. Thus, (3.5) holds for small $\epsilon > 0$ and therefore, by Corollary 3.6, $u(t, \phi^-)$ is monotone nondecreasing in $t \geq 0$. If it is bounded above, then it converges to an equilibrium $v^* > v$. These arguments also apply even when the irreducibility assumption fails provided that the first assertion of Theorem 6.1 holds.

Here, we have constructed a monotone arc of sub-equilibria for the semiflow generated by (1.1) such that hypothesis (U_+) of Proposition 2.5.2 holds. The latter result gives stronger conclusions.

Remark 6.2 is the reaction-diffusion counterpart of Theorem 4.3.3 for ordinary differential equations.

7. The Biochemical Control Circuit with Diffusion

As an application of the ideas of this chapter, we consider the by now familiar example of the biochemical control circuit introduced in section 2 of Chapter 4. It models the control of protein synthesis in the cell but it ignores the diffusion of the proteins in the nucleus and cytoplasm. Here, we take this into account but we will ignore the nucleus and treat the cell as a bounded domain Ω in $\mathbb{R}^3$ ($N = 3$) in which the proteins can diffuse. Let $u_i(x, t)$ denote the concentration of the i-th protein in Ω at time t, $1 \leq i \leq n$. Let d_i be its diffusivity. Then assuming no flux of the constituents out of Ω, the u_i satisfy

$$\frac{\partial u_1}{\partial t}(x,t) = d_1 \Delta u_1(x,t) + g(u_n(x,t)) - \alpha_1 u_1(x,t),$$
$$\frac{\partial u_i}{\partial t}(x,t) = d_i \Delta u_i(x,t) + u_{i-1}(x,t) - \alpha_i u_i(x,t), \quad 2 \leq i \leq n, \quad (7.1)$$
$$0 = \frac{\partial u_i}{\partial \nu}(x,t), \quad x \in \partial \Omega,$$
$$u_i(x,0) = \phi_i(x),$$

The α_i are assumed to be positive constants and g is a bounded, twice continuously differentiable function satisfying

$$M > g(u) > 0, \quad g'(u) > 0, \quad u > 0.$$

Neumann boundary conditions dictate that the appropriate state space for (7.1) is $X = \prod_1^n C(\bar{\Omega})$. As concentrations, the u_i are necessarily nonnegative as are their initial values ϕ_i. Theorem 3.1 and Corollary 3.2 ensure that if $\phi_i \geq 0$ then $u_i(x,t) \geq 0$, $1 \leq i \leq n$. In other words, (7.1) generates a local semiflow on X_+. If we denote by $F(x,u)$ the so-called reaction terms in (7.1), then $F(x,u) = F(u)$ is cooperative and irreducible, as observed in Chapter 4, section 2.

Solutions of the corresponding ordinary differential equation, obtained by ignoring diffusion in (7.1) (see (2.1) of Chapter 4),

$$\begin{aligned} u'_1 &= g(u_n) - \alpha_1 u_1 \\ u'_i &= u_{i-1} - \alpha_i u_i, \quad 2 \leq i \leq n \end{aligned} \quad (7.2)$$

are solutions of (7.1) associated with initial data which are constant functions on Ω. We will use our extensive knowledge about these special solutions, described in section 2 of Chapter 4, to study (7.1). Let E_* and E_{**} denote the smallest and largest equilibrium solutions of (7.2) so $0 \leq E_* \leq E_{**}$. As noted in Chapter 4, E_* attracts the solution of (7.2) corresponding to initial data $u = 0$. If $v^-(t)$ denotes the solution of (7.2) satisfying $v^-(0) = 0$, then

$$v^-(t) \to E_*, \quad t \to \infty.$$

Furthermore (see Chapter 4, section 2), if $w \in \mathbb{R}^n$ is given by

$$w = M(\alpha_1^{-1}, (\alpha_1 \alpha_2)^{-1}, \ldots, (\alpha_1 \alpha_2 \ldots \alpha_n)^{-1})$$

and $v_M^+(t)$ is the solution of (7.2) satisfying $v_M^+(0) = w$, then

$$v_M^+(t) \to E_{**}, \quad t \to \infty, \quad M > 1.$$

Now, if $\phi \in X_+$, we can choose $M > 1$ such that

$$v^-(0) \leq \phi(x) \leq v_M^+(0), \quad x \in \bar{\Omega}.$$

Therefore, by Theorem 3.4, if $u(x,t) = [u(t,\phi)](x)$, we have

$$v^-(t) \leq u(x,t) \leq v_M^+(t), \quad t > 0, x \in \bar{\Omega}. \quad (7.3)$$

Among other things, this shows that the solution $u(t,\phi)$ is defined and bounded for $t \geq 0$ and that (7.1) generates a semiflow on X_+. It is a monotone semiflow by Corollary 3.5 and strongly monotone by Theorem 4.1. Let Λ be the order interval

$[E_*, E_{**}]$ in $\mathbb{R}_+^n$ and X_Λ be the corresponding order interval in X_+. According to (7.3), X_Λ attracts all solutions of (7.1) and, by Corollary 3.3, or by monotonicity, X_Λ is positively invariant for (7.1).

Let E denote the set of equilibria of (7.1). Obviously,
$$E \subset X_\Lambda.$$

The set E consists of solutions v of the boundary value problem

$$\begin{aligned} 0 &= d_1 \Delta v_1(x) + g(v_n(x)) - \alpha_1 v_1(x), \\ 0 &= d_i \Delta v_i(x) + v_{i-1}(x) - \alpha_i v_i(x), \quad 2 \leq i \leq n, \ x \in \Omega \\ 0 &= \frac{\partial v_i}{\partial \nu}(x), \quad x \in \partial\Omega. \end{aligned} \quad (7.4)$$

Of course, any equilibrium of (7.2) is a (constant) equilibrium of (7.1).

The result below should be compared to Proposition 4.2.1. It is our main result concerning (7.1).

THEOREM 7.1. *All solutions of (7.1) are attracted to X_Λ and there is an open and dense subset of X_+ consisting of convergent points for (7.1). If E is a single point $(E_* = E_{**})$, then all solutions converge to it. If E consists of two points then all solutions converge to one of these points.*

PROOF. All solutions of (7.1) are attracted to X_Λ by (7.3). Theorem 5.1 and Theorems 2.3.1 and 2.3.2 establish the remaining assertions. Note that (7.3) implies that (T) holds. □

It is not a trivial matter to determine the set E for (7.1) as it is for (7.2) unless, of course, that $E_* = E_{**}$. The interesting question is whether there are nonconstant equilibria of (7.1). If such equilibria exist and if Ω is convex, then each nonconstant equilibrium is unstable by the result of Kishimoto and Weinberger mentioned in Remark 6.1. The constant equilibria of (6.1), that is, the equilibria of (6.2) have the same stability properties as equilibria of (6.1) as they have as equilibria of (6.2) by Remark 6.1. If Ω is very small (such as in the case of a cell) or if the diffusivities d_i are sufficiently large, then E consists only of constant equilibria and each solution of (6.1) is shadowed by a corresponding solution of (6.2) by a result of Conway, Hoff and Smoller(1978) (see also Smoller(1983)). Thus (6.1) and (6.2) have identical dynamics.

8. Remarks and Discussion

The approach taken in this chapter and most of the results of section 3 on invariance and comparison are strongly influenced by the work of Martin(1976,1978,1979). See also Amann(1978), Fife and Tang(1981), Smoller(1983), and Leung(1989) for related results. The application to the biochemical control loop with diffusion, treated in section 7, has been considered by Martin(1978,1981). Results of Mora(1983) are instrumental for establishing that (1.1) generates a C^1 semiflow on $\prod_1^n C(\bar\Omega)$, in case of Robin or Neumann boundary conditions, and on $\prod_1^n C_0(\bar\Omega)$ and $\prod_1^n C_0^1(\bar\Omega)$, in case of Dirichlet boundary conditions. The various smoothness hypotheses that we assumed were driven by the need to apply these results and to ensure that mild solutions are classical solutions so that maximum principles could be applied.

The results in this chapter can easily be extended to allow F to depend on ∇u, provided suitable bounds on this dependence are satisfied. Furthermore, the boundary conditions need not be all of one type, either Dirichlet or Robin, but can be different for different components of u. See Amann(1978) and Matano(1986,1987) for results in these directions. Of course, the partial order does not need to be the usual one, generated by $\mathbb{R}^n_+$, but could be any of the cones introduced in Chapter 3 as well. Finally, as noted repeatedly in sections 3,4 and 5, we chose $\Lambda = \mathbb{R}^n_+$ merely for convenience and because it is perhaps the canonical example. Any other choice of a closed convex subset Λ for which Theorem 3.1 holds will do.

There is a great deal of important work due to H. Matano concerning monotone methods generally and for parabolic equations in particular. For example, Matano(1978) shows that all bounded solutions of a parabolic equation in one spatial dimension converge to an equilibrium. As mentioned earlier, it was Matano (1984) who first suggested the notion of the strong order preserving property for a semiflow and exploited it in the works Matano(1984,1986,1987). Questions of stability and asymptotic behavior for a single parabolic equation are considered in Matano(1979). A particularly interesting application is to be found in Matano and Mimura(1983). They consider a reaction-diffusion model of two competing species in a bounded spatial region Ω with Neumann boundary conditions and where reaction terms are independent of the spatial variable. If Ω is convex, then the result of Kishimoto and Weinberger implies that nonconstant equilibria are unstable in the linear approximation and therefore, pattern formation cannot occur for convex domains. Matano and Mimura show that spatial pattern formation can occur in certain dumbbell-shaped (nonconvex) domains consisting of two roughly spherical domains joined by a thin connecting strip. Each population dominates its rival in different spherical portions of the domain in an equilibrium solution.

Some of the results in this chapter can be extended to abstract semilinear parabolic evolution equations on certain fractional power spaces corresponding to a sectorial operator. See Poláčik(1989a) and Smith and Thieme(1991a). Most of the results of this chapter can be extended as well to reaction-diffusion systems containing time delays in the "reaction terms". See Martin and Smith(1990,1991).

As noted in previous chapters, Theorem 6.1 is essentially a special case of abstract results for positive semigroups. See Kerscher and Nagel(1984). The book of Clements et al(1987) includes a self-contained treatment of these results.

It is possible to construct quasimonotone comparison systems relative to a not necessarily quasimonotone system possessing an invariant rectangle as we did in Chapter 5, section 2 for delay differential systems. Furthermore, the method of contracting rectangles can be applied. The interested reader is referred to Smoller(1983) or Fife(1979).

Finally, it should be mentioned that a very influential work on monotonicity methods for partial differential equations is Sattinger(1972). Many of the ideas discussed here were described in that work. For other works employing monotonicity methods to partial differential equations, see the texts Fife(1979), Smoller(1983), Leung(1989) and Pao(1992).

For a fairly complete analysis of the asymptotic behavior of a scalar semilinear reaction diffusion equation with reaction term that does not grow too fast at infinity, see Lions(1984). In the scalar case, when the differential operator is self-adjoint, not only is the semiflow strongly monotone but there is a Liapunov function which decreases along solutions.

Quite general results on convergence to equilibrium and stability of solutions of weakly coupled, cooperative parabolic systems appear in the work of Vischnevskiĭ (1992).

CHAPTER 8

A Competition Model

0. Introduction

In this concluding chapter, the results of the Chapter 7 are applied to a model of microbial competition for a nutrient in a tubular reactor that was formulated by Kung and Baltzis(1992). The model is introduced in the next section. Its analysis requires the application of most of the results of the previous chapter. In particular, as the model system is reduced to a competitive reaction-diffusion system consisting of a pair of equations, we immediately require the application of the monotonicity results of the previous chapter adapted to the partial order characteristic of two-dimensional competitive systems, namely that generated by the fourth quadrant in the plane. Furthermore, as is typically the case in applications, the irreducibility assumption holds only in the interior of the state-space and therefore, Theorem 7.5.1 implying generic convergence to equilibrium cannot be immediately applied. As a consequence, comparison methods and differential inequality arguments play a large role in the analysis.

We begin by describing the model setup.

1. The Model

Consider a thin tubular culture vessel of length L in which two microbial populations u_1 and u_2 compete for nutrient S. The competition is purely exploitative: the organisms simply consume the nutrient, thereby making it unavailable for a competitor. We let the symbols above denote the concentration per unit length of these quantities. Assume that the culture vessel, occupying the interval $0 < x < L$ along the x-axis, is fed with growth medium at a constant rate at $x = 0$ due to a constant laminar flow velocity v of the fluid in the vessel in the direction of increasing x. This external feed contains nutrient at concentration S^0. Medium, nutrient and organisms exit the vessel at $x = L$ with velocity v. Nutrient and organisms are assumed to diffuse in the vessel with diffusivities which do not differ significantly. Thus, we assume equal diffusivities d for both organisms and the nutrient. See **Figure 1.1** for a schematic description of the reactor. The equations describing the concentrations S, u_1, u_2 are then given below.

$$\begin{aligned}
\frac{\partial S}{\partial t} &= d\frac{\partial^2 S}{\partial x^2} - v\frac{\partial S}{\partial x} - \gamma_1^{-1} u_1 f_1(S) - \gamma_2^{-1} u_2 f_2(S) \\
\frac{\partial u_1}{\partial t} &= d\frac{\partial^2 u_1}{\partial x^2} - v\frac{\partial u_1}{\partial x} + u_1 f_1(S) \\
\frac{\partial u_2}{\partial t} &= d\frac{\partial^2 u_2}{\partial x^2} - v\frac{\partial u_2}{\partial x} + u_2 f_2(S), \quad 0 < x < L, \quad t > 0.
\end{aligned} \quad (1.1)$$

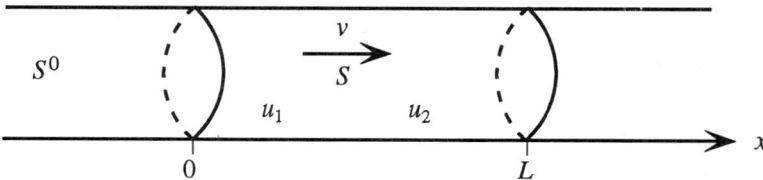

FIGURE 1.1. The tubular culture vessel. Laminar flow of medium with velocity v brings fresh nutrient at concentration S^0 into the vessel.

$$d\frac{\partial S}{\partial x}(0,t) - vS(0,t) = -vS^0$$
$$d\frac{\partial u_i}{\partial x}(0,t) - vu_i(0,t) = 0$$
$$\frac{\partial S}{\partial x}(L,t) = \frac{\partial u_i}{\partial x}(L,t) = 0, \quad i = 1,2. \tag{1.2}$$

$$S(x,0) = S_0(x) \geq 0, \ u_i(x,0) = u_{i0}(x) \geq 0, \ 0 \leq x \leq L. \tag{1.3}$$

The nonlinear functions $f_i(S)$ describe the growth rate of the i-th organism at nutrient concentration S. We assume these functions satisfy

$$f_i(0) = 0, \quad f_i'(S) > 0, \quad f_i \in C^2.$$

A classical example is the Monod function

$$f_i(S) = \frac{m_i S}{a_i + S}, \quad m_i, a_i > 0.$$

The parameters m_i and a_i are, respectively, the maximum growth rate and the half-saturation constant. The latter is the nutrient concentration at which growth is half maximum.

As a check that the boundary conditions are appropriate, we integrate the equation for S from $x = 0$ to $x = L$, obtaining

$$\frac{d}{dt}\int_0^L S(x,t)dx = v(S^0 - S(L,t)) - \sum_i \gamma_i^{-1}\int_0^L u_i(x,t)f_i(S(x,t))dx.$$

Thus, the rate of change of nutrient in the vessel is the difference between the inflow rate at $x = 0$ and the outflow rate at $x = L$ minus consumption of nutrient by the two populations of organisms. The reader is invited to perform a similar integration of the equation for u_i to verify that the rate of change of organism u_i in the vessel is the difference between the natural growth of the population and washout at $x = L$.

The equations can be simplified by non-dimensionalizing the parameters, dependent and independent variables. Nondimensional quantities are indicated below with bars.

$$\bar{S} = S/S^0, \ \bar{u}_i = u_i/\gamma_i S^0, \ \bar{x} = x/L, \ \bar{t} = vt/L, \ \bar{d} = d/Lv, \ \bar{f}_i(\bar{S}) = (L/v)f_i(S^0\bar{S}).$$

1. THE MODEL

In order to conserve notation, we drop the bars over the nondimensional quantities, returning to the original notation.

The boundary condition for the nutrient is nonhomogeneous and consequently the system (1.1)-(1.3) is not of the form considered in the previous chapter. However, the change of variables $N = S - S^0$ in (1.1)-(1.3) removes the inhomogeneous term in the boundary conditions. In terms of the scaled quantities, this change of variables is
$$N = S - 1.$$
With the scaling and change of variables as described above, the equation (1.1) becomes the following.

$$\begin{aligned}
\frac{\partial N}{\partial t} &= d\frac{\partial^2 N}{\partial x^2} - \frac{\partial N}{\partial x} - u_1 f_1(1+N) - u_2 f_2(1+N) \\
\frac{\partial u_1}{\partial t} &= d\frac{\partial^2 u_1}{\partial x^2} - \frac{\partial u_1}{\partial x} + u_1 f_1(1+N) \\
\frac{\partial u_2}{\partial t} &= d\frac{\partial^2 u_2}{\partial x^2} - \frac{\partial u_2}{\partial x} + u_2 f_2(1+N), \quad 0 < x < 1, \quad t > 0.
\end{aligned} \quad (1.4)$$

The boundary conditions are now homogeneous.

$$\begin{aligned}
d\frac{\partial N}{\partial x}(0,t) - N(0,t) &= 0 \\
d\frac{\partial u_i}{\partial x}(0,t) - u_i(0,t) &= 0 \\
\frac{\partial N}{\partial x}(1,t) = \frac{\partial u_i}{\partial x}(1,t) &= 0.
\end{aligned} \quad (1.5)$$

Initial data are given by
$$N(x,0) = N_0(x), \; u_i(x,0) = u_{i0}(x), \; 0 \le x \le 1. \quad (1.6)$$

Since $S = 1 + N$ must be nonnegative, the biologically relevant region for the variables N, u_1, u_2 is
$$\Lambda = \{(N, u_1, u_2) \in \mathbb{R}^3 : N \ge -1, u_i \ge 0, i = 1, 2\}.$$

Let $X = \prod_1^3 C([0,1], \mathbb{R})$ and X_Λ be the set of those functions in X taking values in Λ. It will be shown below that (1.4)-(1.6) generates a semiflow on X_Λ. To accomplish this, we must show that solutions can be extended to $t \ge 0$. In fact, solutions are bounded as one can see by examining the equation satisfied by the total nutrient concentration consisting of the "raw" nutrient S plus "stored" nutrient $u_1 + u_2$. That is, set
$$B = N + u_1 + u_2.$$
Then B satisfies the linear equation
$$\frac{\partial B}{\partial t} = d\frac{\partial^2 B}{\partial x^2} - \frac{\partial B}{\partial x}$$
$$d\frac{\partial B}{\partial x}(0,t) - B(0,t) = 0 = \frac{\partial B}{\partial x}(1,t).$$
The asymptotic behavior of solutions of this linear equation is determined by the solutions of the eigenvalue problem
$$\lambda w = d\frac{d^2 w}{dx^2} - \frac{dw}{dx} \quad (1.7)$$

$$d\frac{dw}{dx}(0) - w(0) = 0 = \frac{dw}{dx}(1).$$

There are countably many eigenvalues λ_n, $n \geq 0$, which are all real and satisfy $\lambda_n \to -\infty$ as $n \to \infty$. They may be ordered as follows

$$\ldots < \lambda_{n+1} < \lambda_n < \ldots < \lambda_1 < \lambda_0.$$

The estimate

$$\lambda_0 < -\frac{1}{1+d}$$

is easily established but the essential point is that $\lambda_0 < 0$. Indeed, if Q denotes the differential operator on the right in (1.7), then an integration by parts shows that

$$\int_0^1 (Qu)(x)u(x)dx = -\frac{1}{2}(u(0)^2 + u(1)^2) - d\int_0^1 \left[\frac{du}{dx}(x)\right]^2 dx < 0$$

if u satisfies the boundary conditions. Setting $u = w_0$ in the inequality, where w_0 is the principal eigenvector of (1.7) ($Qw_0 = \lambda_0 w_0$), leads to the conclusion $\lambda_0 < 0$. The sharper estimate on λ_0 follows on putting $w = d^{-1}x + 1$ into Theorem 15, p.38, (6) of Protter and Weinberger (1967).

Thus the principal eigenvalue of (1.7) is negative and the corresponding principal eigenfunction satisfies $w_0(x) > 0$ for all $x \in [0,1]$ by Theorem 7.6.1. As a consequence, $B(x,t)$ must tend to zero as t increases.

PROPOSITION 1.1. *For each $(N_0, u_{10}, u_{20}) \in X_\Lambda$, there exists a unique solution of (1.4)-(1.6) defined for all $t \geq 0$ and this solution remains in X_Λ. Moreover,*

$$N(x,t) + u_1(x,t) + u_2(x,t) \to 0, \quad t \to \infty$$

uniformly for $0 \leq x \leq 1$. Therefore, solutions of (1.4)-(1.6) are bounded.

PROOF. The existence of a solution in X_Λ follows from an application of Corollary 7.3.3. As noted above, the principal eigenvalue of the eigenvalue problem (1.7) satisfies $\lambda_0 < 0$ and the corresponding eigenfunction $w_0(x) > 0$ for all $x \in [0,1]$. Choose α and β such that $\alpha w_0(x) \leq B(x,0) \leq \beta w_0(x)$ holds for all $x \in [0,1]$, where $B(x,0) = N(x,0) + u_1(x,0) + u_2(x,0)$. Then, by Corollary 7.2.3,

$$\alpha e^{\lambda_0 t} w_0(x) \leq B(x,t) \leq \beta e^{\lambda_0 t} w_0(x)$$

for $0 \leq x \leq 1$ and $t > 0$. This establishes the boundedness of solutions. □

Proposition 1.1 establishes that $N + u_1 + u_2$ converges to zero at an exponential rate as t increases. Biologically, this is just the statement that the total nutrient must asymptotically approach the input concentration from the reservoir. Thus, asymptotically, we have that

$$N = -u_1 - u_2.$$

Certainly, this relationship holds on the omega limit set of any solution of (1.4)-(1.6). Consequently, as a first step, it is reasonable to pass to the limiting system where this relation holds. The limiting system is obtained by dropping the equation for N and using the relation above in (1.4)-(1.6) to get

$$\frac{\partial u_i}{\partial t} = d\frac{\partial^2 u_i}{\partial x^2} - \frac{\partial u_i}{\partial x} + u_i f_i(1 - u_1 - u_2), \quad 0 < x < 1, \quad t > 0. \tag{1.8}$$

1. THE MODEL

$$d\frac{\partial u_i}{\partial x}(0,t) - u_i(0,t) = \frac{\partial u_i}{\partial x}(1,t) = 0. \tag{1.9}$$

$$u_i(x,0) = u_{i0}(x), \ 0 \leq x \leq 1, \ i = 1,2. \tag{1.10}$$

Since $N \geq -1$, the biologically relevant region for the limiting system is given by

$$\Theta = \{(u_1, u_2) \in \mathbb{R}^2 : u_i \geq 0, u_1 + u_2 \leq 1\}.$$

For the remainder of this chapter, we focus attention on the system (1.8)-(1.10). In the discussion and remarks section, we will comment on the relationship between our results for it and the dynamics of the original system (1.4)-(1.6). We begin our study of the limiting system by noting that it is well-posed and by obtaining some elementary properties of its solutions. Hereafter, except for section 2, X denotes the space $\prod_1^2 C([0,1], \mathbb{R})$ and X_Θ, the subset of X consisting of functions taking values in Θ.

PROPOSITION 1.2. *For each $(u_{10}, u_{20}) \in X_\Theta$, there exists a unique solution of (1.8)-(1.10) defined for $t \geq 0$ and this solution remains in X_Θ. Moreover, if $u_{i0}(x) > 0$ for some $x \in [0,1]$, then $u_i(x,t) > 0$ for all $x \in [0,1]$ and $t > 0$.*

PROOF. Except for the last assertion, we can apply Theorem 7.3.1. Hypotheses (3.1) and (3.2) of that result must be verified. (3.2) requires that $T(t)X_\Theta \subset X_\Theta$ for $t \geq 0$. Now, $T(t) = (T_1(t), T_1(t))$ where $T_1(t)$ is the semigroup generated by the linear equation above (1.7) on $C([0,1], \mathbb{R})$. It is a positive semigroup by Corollary 7.2.3 and it leaves positively invariant the subset of functions taking values in $[0,1]$, by Corollary 7.2.4. Therefore, if $\phi = (\phi_1, \phi_2) \in X_\Theta$, then $T(t)\phi = (T_1(t)\phi_1, T_1(t)\phi_2)$ satisfies $T_1(t)\phi_i \geq 0$ and $0 \leq T_1(t)\phi_1 + T_1(t)\phi_2 = T_1(t)(\phi_1 + \phi_2) \leq 1$. The last set of inequalities follows from the fact that $0 \leq \phi_1 + \phi_2 \leq 1$. Thus (3.2) of the previous chapter holds. In order to verify (3.1), note that $F(v) = (v_1 f_1(1 - v_1 - v_2), v_2 f_2(1 - v_1 - v_2))$ for $v = (v_1, v_2) \in \Theta$. It suffices to verify (3.1) for $v \in \partial\Theta$. Consider the case that $v \in \Theta$, $v_i > 0$, and $v_1 + v_2 = 1$. Then $v + hF(v) = v$ since $F(v) = 0$ so clearly, (3.1) holds for such v. In a similar fashion, (3.1) is verified on the other parts of the boundary. Thus the hypotheses of Theorem 7.3.1 are verified. The first assertions of the proposition follow immediately from this result and the fact that X_Θ is bounded, implying that solutions can be extended to $t \geq 0$.

Now suppose that $u_{i0}(x) > 0$ for some x. Then $v(x,t) = -u_i(x,t)$ satisfies $v(x,t) \leq 0$ and the differential inequality

$$d\frac{\partial^2 v}{\partial x^2} - \frac{\partial v}{\partial x} - \frac{\partial v}{\partial t} \geq 0.$$

By Theorem 7.2.1, if $v(x,t) = 0$ for some (x_0, t_0) with $x_0 \in (0,1)$ and $t_0 > 0$, then $v(x,t) = 0$ for all (x,t) with $t \leq t_0$ and $x \in [0,1]$. This would contradict that $v(x,0) < 0$ for some x. Therefore, v cannot assume the value zero except possibly for $(0, t_0)$ or $(1, t_0)$ for some $t_0 > 0$. If $v(1, t_0) = 0$, then $\frac{\partial v}{\partial x}(1, t_0) > 0$ by Theorem 7.2.2. But this would contradict the fact that v satisfies the boundary conditions (1.12). A similar argument shows that $v(0,t)$ cannot vanish for $t > 0$. Thus $v(x,t) < 0$ for $x \in [0,1]$ and $t > 0$. This proves the last assertion. □

The last assertion of Proposition 1.2 has important consequences. It says, with slight abuse of notation, that if initially the organism u_i is present somewhere (for some x) in the culture vessel, then subsequently, it is present throughout the vessel. In particular, if (u_1, u_2) is an equilibrium solution of (1.8)-(1.10), then for each i, either $u_i(x) > 0$ for all $x \in [0,1]$ or $u_i(x) \equiv 0$ in $[0,1]$. A population is present everywhere or nowhere in the vessel at equilibrium. Equilibrium solutions of (1.8)-(1.10) satisfy

$$0 = d\frac{d^2 u_i}{dx^2} - \frac{du_i}{dx} + u_i f_i(1 - u_1 - u_2), \quad 0 < x < 1. \tag{1.11}$$

$$d\frac{du_i}{dx}(0) - u_i(0) = \frac{du_i}{dx}(1) = 0. \tag{1.12}$$

By virtue of Proposition 1.2, the limiting system induces a semiflow Φ on X_Θ given by

$$\Phi_t(u_0) = u(t)$$

where $u_0 = (u_{10}, u_{20})$ and $u(t) = (u_1(\bullet, t), u_2(\bullet, t))$ is the corresponding solution of (1.8)-(1.10). It is this semiflow which we study in this chapter.

The limiting system is obviously a competitive system and since it is two dimensional, it can be viewed as a monotone system with respect to the cone K where K is the fourth quadrant:

$$K = \{(u_1, u_2) \in X : u_1 \geq 0, u_2 \leq 0\}.$$

As usual, K generates a partial order on X as follows

$$(u_1, u_2) \leq_K (\bar{u}_1, \bar{u}_2) \iff u_1 \leq \bar{u}_1 \text{ and } u_2 \geq \bar{u}_2.$$

We also use the notation $<_K$ and $\ll_K$ with their usual meaning. The next result summarizes the monotonicity properties of the semiflow Φ.

PROPOSITION 1.3. *The semiflow Φ on X_Θ is monotone with respect to the order $\leq_K$ and it is strongly monotone on X_Θ in the following conditional sense: if $(u_1, u_2) <_K (\bar{u}_1, \bar{u}_2)$ and either $0 < u_i$ for $i = 1,2$ or $0 < \bar{u}_i$ for $i = 1,2$, then $\Phi_t((u_1, u_2)) \ll_K \Phi_t((\bar{u}_1, \bar{u}_2))$ for $t > 0$.*

PROOF. We will apply Corollary 7.3.5 and Theorem 7.4.1 to the transformed system obtained from (1.8)-(1.10) by the change of dependent variables $v_1 = u_1$ and $v_2 = -u_2$. In these variables, the system becomes

$$\frac{\partial v_i}{\partial t} = d\frac{\partial^2 v_i}{\partial x^2} - \frac{\partial v_i}{\partial x} + v_i f_i(1 + v_2 - v_1)$$

with the usual boundary conditions. The domain Θ is transformed to the new domain $\Theta' = \{(v_1, v_2) \in \mathbb{R}^2 : v_1 \geq 0, v_2 \leq 0, v_1 \leq 1 + v_2\}$. It is easy to check that $F : \Theta' \to \mathbb{R}^2$ defined by $F(v) = (v_1 f_1(1 + v_2 - v_1), v_2 f_2(1 + v_2 - v_1))$ is cooperative on Θ' and cooperative and irreducible on that part of Θ' for which $v_i \neq 0$, $i = 1, 2$. Therefore, the system above with the usual boundary conditions generates a monotone semiflow on $X_{\Theta'}$ with respect to the usual ordering induced by the cone X_+. Interpreting this result in terms of the original system (1.8)-(1.10) proves that Φ is monotone with respect to the ordering $\leq_K$ on X_Θ.

The conditional strong monotonicity follows from the proof of Theorem 7.4.1 applied to the transformed system above together with the fact that if, say $0 < u_i$ for $i = 1, 2$, then $u_i(x, t) > 0$ for all (x, t) with $t > 0$ and $i = 1, 2$, where $(u_1(x,t), u_2(x,t))$ is the corresponding solution of (1.8)-(1.10). The latter was established in Proposition 1.2. This ensures that in the proof of Theorem 7.4.1, the

2. A Single Population

It is useful to consider first the case that there is only a single population in the culture. Mathematically, this just means that either u_2 or u_1 is set to zero in the equations (1.8)-(1.10). In order to simplify notation, all subscripts are dropped in the remaining equations and we consider

$$\frac{\partial u}{\partial t} = d\frac{\partial^2 u}{\partial x^2} - \frac{\partial u}{\partial x} + uf(1-u) \tag{2.1}$$

with boundary conditions given by

$$d\frac{\partial u}{\partial x}(0,t) - u(0,t) = \frac{\partial u}{\partial x}(1,t) = 0 \tag{2.2}$$

and initial conditions given by

$$u(x,0) = u_0(x). \tag{2.3}$$

For this section only, let $X = C([0,1], \mathbb{R})$ and consider it as the appropriate state space for (2.1)-(2.3). Of course, the biologically relevant domain is $X_{[0,1]}$ since u must satisfy $0 \leq u \leq 1$.

It turns out that the dynamics of (2.1)-(2.3) is entirely determined by the stability properties of the trivial solution $u = 0$. The variational system corresponding to this equilibrium is given by

$$\frac{\partial v}{\partial t} = d\frac{\partial^2 v}{\partial x^2} - \frac{\partial v}{\partial x} + vf(1) \tag{2.4}$$

$$d\frac{\partial v}{\partial x}(0,t) - v(0,t) = \frac{\partial v}{\partial x}(1,t) = 0. \tag{2.5}$$

The stability properties of the trivial solution are therefore governed by the solutions of the eigenvalue problem

$$\lambda w = d\frac{d^2 w}{dx^2} - \frac{dw}{dx} + wf(1) \tag{2.6}$$

$$d\frac{dw}{dx}(0) - w(0) = \frac{dw}{dx}(1) = 0. \tag{2.7}$$

Comparing this eigenvalue problem with (1.7), which has eigenvalues λ_n, we find that the eigenvalues of (2.6)-(2.7) are given by $\lambda_n + f(1)$ for $n \geq 0$. In particular, corresponding to the principal eigenvalue $\lambda_0 + f(1)$, the principal eigenvector is w_0, the principal eigenvector for (1.7). The next result states that the chemostat is inhospitable to the organism if the principal eigenvalue is negative and provides a satisfactory environment for the organism if it is positive. In the latter case, the population density tends to an positive equilibrium which is a monotonically increasing function of $x \in [0,1]$.

THEOREM 2.1. *For each $u_0 \in X_{[0,1]}$, there exists a unique solution of (2.1)-(2.3) defined for $t \geq 0$ and it remains in $X_{[0,1]}$. If $\lambda_0 + f(1) < 0$, then*

$$u(x,t) \to 0, \quad t \to \infty$$

uniformly for $x \in [0,1]$. If $\lambda_0 + f(1) > 0$, then there exists a unique nontrivial equilibrium U in $X_{[0,1]}$ and it satisfies $0 < U(x) < 1$ for $x \in [0,1]$, $U'(x) > 0$ for $x \in [0,1)$, and

$$u(x,t) \to U(x), \quad t \to \infty$$

uniformly for $x \in [0,1]$ provided $u_0(x) > 0$ for some $x \in [0,1]$.

PROOF. The first assertion concerning the existence of a solution of (2.1)-(2.3) remaining in $X_{[0,1]}$ follows from Proposition 1.2 and the fact that if $u_{i0} = 0$ then $u_i(x,t) \equiv 0$. Let $\mu = \lambda_0 + f(1)$ and suppose first that $\mu < 0$. As

$$0 \leq uf(1-u) \leq f(1)u, \quad 0 \leq u \leq 1,$$

we can apply Theorem 7.3.4 with $F^- = 0$ and $F^+ = f(1)u$. Letting $v^- = 0 \leq u_0 \leq v^+ = \alpha w_0$, for suitable $\alpha > 0$, we may conclude from that result that $0 \leq u(x,t) \leq \alpha w_0(x)e^{\mu t}$. Since $\mu < 0$, this shows that $u(x,t) \to 0$ as $t \to \infty$ uniformly in $x \in [0,1]$.

Suppose that $\mu > 0$. First note that (2.1)-(2.3) generates a strongly monotone semiflow Φ on $X_{[0,1]}$ by Theorem 7.4.1. By Remark 6.2 of Chapter 7, there is a monotone increasing solution of (2.1)-(2.3) initiating from $u_0 = \epsilon w_0$ for all small positive ϵ. Since this solution remains in $X_{[0,1]}$, it is bounded and therefore must converge to an equilibrium (see the proof of Corollary 7.3.6). Therefore, the existence of a positive equilibrium is established. We show the uniqueness of this equilibrium below. Having shown that there are only two equilibria, 0 and U, it follows that all solutions converge to one of these by Theorem 2.3.2. Except for the trivial solution, all others must converge to the positive equilibrium U. Indeed, if $u_0 \neq 0$, then $u(x,t) > 0$ for all (x,t) with $t > 0$ and $0 \leq x \leq 1$, by Proposition 1.2. Thus, there exists $\epsilon > 0$ such that $\epsilon w_0 \leq u(\bullet, 1)$ and by monotonicity of the semiflow, $\Phi_t(\epsilon w_0) \leq \Phi_t(u(\bullet, 1)) = u(\bullet, 1+t)$. Since $\Phi_t(\epsilon w_0)$ converges monotonically to U as $t \to \infty$, if ϵ is sufficiently small, it follows that $u(\bullet, t) \to U$ as $t \to \infty$.

Before establishing the uniqueness of the positive equilibrium, we note that the positive equilibrium U whose existence was established in the previous paragraph satisfies $\epsilon w_0(x) < U(x) < 1$. To see this, apply Corollary 3.6 with $\phi^+ \equiv 1$ and observe that (3.5) of Chapter 7 holds with inequalities reversed. Consequently, the solution $u(x,t)$ of (2.1)-(2.3) with $u_0 = 1$ is monotone decreasing in t and converges to an equilibrium, by the above mentioned result. More precisely, if $\epsilon > 0$ is such that $\epsilon w_0(x) < 1$ for all $x \in [0,1]$ and such that $\Phi_t(\epsilon w_0)$ increases monotonically, then the strong monotonicity of Φ implies that for each $t > 0$

$$\epsilon w_0 \ll \Phi_t(\epsilon w_0) \ll \Phi_t(1) \ll 1$$

where 1 denotes the function that is identically one. Since both solutions converge, there is an equilibrium U satisfying $\epsilon w_0 \ll U \ll 1$.

It remains to establish the uniqueness of the nonzero equilibrium of (2.1) and (2.2). An equilibrium must satisfy

$$-d\frac{d^2 U}{dx^2} + \frac{dU}{dx} + \alpha^2 U = U[\alpha^2 + f(1-U)] \qquad (2.8)$$

on $0 \leq x \leq 1$ and the boundary conditions. It is convenient to extend the definition of f to $u \leq 0$ so that $f'(u) \geq 0$ for all $u \leq 1$ and so that f is a (negative) constant on the interval $(-\infty, -1]$. With this extension, $F(u) = u[\alpha^2 + f(1-u)]$ is defined for all $u \geq 0$ and we may choose $\alpha > 0$ so large that $[\alpha^2 + f(1-u)] > 0$ for $u \geq 0$ and such that F has a positive derivative for $u \geq 0$.

The differential operator on the left side of (2.8), together with the boundary conditions, has a strongly positive, compact inverse $L : C([0,1], \mathbb{R}) \to C([0,1], \mathbb{R})$. See Amann(1976), Theorem 4.2. Therefore, an equilibrium must be a fixed point of the operator equation
$$S(U) = L(F(U(\bullet)))$$
in the cone $X_+ = C([0,1], \mathbb{R}_+)$. Observe that $S(0) = 0$ and that if $0 < U_1 \leq U_2$, then $0 \ll S(U_1) \leq S(U_2)$ since F is strictly increasing and L is strongly positive. Furthermore, S is concave in the sense that if $U \gg 0$ and $0 < \eta < 1$, then there exists $\beta = \beta(\eta, U) > 0$ such that $S(\eta U) \geq (1+\beta)\eta S(U)$. It then follows from Theorem 6.3 of Krasnoselskii(1964) that S has a unique nontrivial fixed point in X_+. As this argument is brief, we give it here. If there were two distinct positive equilibria U_i, $i = 1, 2$, then one of the inequalities $U_1 \leq U_2$ or $U_2 \leq U_1$ must fail to hold. For definiteness, suppose the former does not hold. Since $0 \ll U_2$, there exists $\eta > 0$ such that $\eta U_1 \leq U_2$ and η is maximal with this property ($\eta U_1(x_0) = U_2(x_0)$ for some x_0). Since $U_1 \leq U_2$ does not hold, it follows that $\eta < 1$. But then the concavity of S implies that
$$U_2 = S(U_2) \geq S(\eta U_1) \geq (1+\beta)\eta S(U_1) = (1+\beta)\eta U_1$$
which contradicts the maximality of η.

Finally, we show that $U'(x) > 0$ for $x \in [0,1)$. First note that since $U(0) > 0$ it follows from the boundary conditions that $dU'(0) = U(0) > 0$. If $U'(x) = 0$ for some $x < 1$ let x_0 be the smallest such x. Since $U''(x_0) < 0$ follows from the equation, $U(x_0)$ is a strict local maximum of U. Integrating the equation satisfied by U from x to 1 leads to
$$dU'(x) = U(x) - U(1) + \int_x^1 Uf(1-U)dx$$
From this we see that $U(1) > U(x_0)$ and therefore U must have a local minimum at a point $x_1 \in (x_0, 1)$. But at such a point, $U'(x_1) = 0$ and $U''(x_1) < 0$, a contradiction. Therefore, $U'(x) > 0$ for $x \in [0,1)$. □

Theorem 2.1 can be applied to either of the two systems obtained from (1.8)-(1.10) by setting one of the two u_i to zero. Therefore, we conclude that (1.8)-(1.10) has the following equilibria in case the given condition is satisfied.
$$E_0 = (0,0)$$
$$E_1 = (U_1, 0) \text{ exists} \iff \lambda_0 + f_1(1) > 0.$$
$$E_2 = (0, U_2) \text{ exists} \iff \lambda_0 + f_2(1) > 0.$$
Here, U_i denotes the positive equilibrium of (2.1),(2.2) resulting from putting $f = f_i$. Of course, there may be additional equilibria as well and these must be positive, as noted in the previous section. If a positive equilibrium exists, then the two organisms can, in principle, coexist in the culture vessel although coexistence would only be observable if such an equilibrium were stable.

3. Stability of the Equilibria E_0, E_1, E_2.

In this section, the stability properties of the equilibria E_i, $i = 0, 1, 2$ are examined. We begin with the trivial equilibrium, obtaining conditions for stability and instability. It turns out that local stability implies global stability for E_0.

THEOREM 3.1. *If for some i, $\lambda_0 + f_i(1) < 0$, then*

$$u_i(x, t) \to 0, \quad t \to \infty$$

uniformly in $x \in [0, 1]$ for every solution of (1.8)-(1.10). If $\lambda_0 + f_i(1) < 0$ for $i = 1, 2$, then E_0 attracts all solutions of (1.6)-(1.8). If $\lambda_0 + f_i(1) > 0$ for $i = 1, 2$ then there exists a neighborhood U of E_0 in X_Θ such that for each $u \in U$ distinct from E_0, there is a $\tau = \tau(u) > 0$ for which

$$\Phi_\tau(u) \in \partial U.$$

In other words, E_0 is a repeller in X_Θ.

PROOF. The proof of the first assertion is exactly the same as the proof that $u \to 0$ as $t \to \infty$ in Theorem 2.1, since each u_i can be treated separately. The second assertion follows directly from the first. It remains to show that E_0 is a repeller.

Choose $h \in (0, 1)$ such that $\lambda_0 + f_i(1 - h) > 0$ for $i = 1, 2$ and let $U = \{u = (u_1, u_2) \in X_\Theta : u_1(x) + u_2(x) < h, 0 \le x \le 1\}$. If $u_0 = (u_{10}, u_{20}) \in U$ and $u(t)$ is the corresponding solution of (1.8)-(1.10), then $f_i(1 - u_1(t) - u_2(t)) \ge f_i(1 - h)$ so long as $u(t)$ remains in $\bar{U}$. Consequently, the i-th component of $u(t)$ satisfies

$$\frac{\partial u_i}{\partial t} \ge d\frac{\partial^2 u_i}{\partial x^2} - \frac{\partial u_i}{\partial x} + u_i f_i(1 - h) \tag{3.1}$$

so long as $u(t) \in \bar{U}$. Since $u_0 \ne E_0$, it follows that $u_i(t) \gg 0$ for one or both i by Proposition 1.2 and therefore we may as well assume that $u_{0i} \gg 0$ for some or both i. Then there exists $s > 0$ such that $u_{i0} \ge sw_0$ for at least one i. For definiteness, suppose that $u_{10} \ge sw_0$. As $v_1(x, t) = se^{ct}w_0(x)$ satisfies the linear differential equality corresponding to the differential inequality above, $v(x, 0) = sw_0(x)$ and v satisfies the boundary conditions, with $c = \lambda_0 + f_1(1 - h) > 0$, it follows from Theorem 7.3.4 (with $F^+ = F = f_1(1 - h)u_1$) that $u_1(t) \ge se^{ct}w_0$ so long as $u(t)$ remains in $\bar{U}$. Clearly, this inequality cannot hold for all $t \ge 0$, proving the final assertion. □

Now consider the equilibrium $E_1 = (U_1, 0)$, assuming that $\lambda_0 + f_1(1) > 0$. Its stability is governed by the variational equation given by

$$\begin{aligned}\frac{\partial v_1}{\partial t} &= d\frac{\partial^2 v_1}{\partial x^2} - \frac{\partial v_1}{\partial x} + v_1[f_1(1 - U_1) - U_1 f_1'(1 - U_1)] - v_2 f_1'(1 - U_1)U_1 \\ \frac{\partial v_2}{\partial t} &= d\frac{\partial^2 v_2}{\partial x^2} - \frac{\partial v_2}{\partial x} + v_2 f_2(1 - U_1), \quad 0 < x < 1, \quad t > 0.\end{aligned} \tag{3.2}$$

The corresponding eigenvalue problem is

$$\begin{aligned}\lambda w_1 &= d\frac{d^2 w_1}{dx^2} - \frac{dw_1}{dx} + w_1[f_1(1 - U_1) - U_1 f_1'(1 - U_1)] - w_2 f_1'(1 - U_1)U_1 \\ \lambda w_2 &= d\frac{d^2 w_2}{dx^2} - \frac{dw_2}{dx} + w_2 f_2(1 - U_1), \quad 0 < x < 1.\end{aligned} \tag{3.3}$$

with the usual boundary condition. The significant feature about this eigenvalue problem is that the second equation is decoupled from the first. Let Λ_1 denote the principal eigenvalue of the eigenvalue problem

$$\lambda w = d\frac{d^2 w}{dx^2} - \frac{dw}{dx} + w f_2(1 - U_1), \quad 0 < x < 1, \tag{3.4}$$

together with the usual boundary condition. The stability properties of E_1 are determined by the sign of Λ_1 as the next result makes clear. Here, we use Theorem 2.5.3 for the first time so the reader may wish to review section 5 of Chapter 2.

THEOREM 3.2. *Suppose that $\lambda_0 + f_1(1) > 0$ so E_1 exists. Then E_1 is asymptotically stable if $\Lambda_1 < 0$ and unstable if $\Lambda_1 > 0$. If the latter holds, then $\lambda_0 + f_2(1) > 0$ so E_2 must also exist. Furthermore, if $\Lambda_1 > 0$, then there is a full orbit $\phi : \mathbb{R} \to X_\Omega$ of the semiflow Φ generated by (1.8), (1.9) satisfying:*
(i) *$E_2 \ll_K \phi(t_2) \ll_K \phi(t_1) \ll_K E_1$ if $t_1 < t_2$.*
(ii) *$\phi(t) \to E_1, \quad t \to -\infty$.*
(iii) *There exists an equilibrium E_* of (1.8),(1.9), possibly identical to E_2, satisfying $E_2 \leq_K E_* \ll_K E_1$ and*

$$\phi(t) \to E_*, \quad t \to +\infty.$$

Furthermore, E_* attracts all solutions of (1.8)-(1.10) for which

$$E_* \leq_K (u_{10}, u_{20}) \leq_K E_1 \quad \text{and } u_{20} > 0. \tag{3.5}$$

PROOF. The form of (3.3) implies that if λ is an eigenvalue then either λ is an eigenvalue of (3.4) or $w_2 = 0$ in (3.3) in which case λ is an eigenvalue of the resulting scalar eigenvalue problem with eigenvector $(w_1, 0)$. In the paragraph below it is shown that in the latter case, $\lambda < 0$. Therefore, the eigenvalues of (3.3) are real and there is a nonnegative eigenvalue if and only if $\Lambda_1 \geq 0$. By Theorem 7.6.2, E_1 is asymptotically stable if $\Lambda_1 < 0$ and unstable if $\Lambda_1 > 0$.

Because U_1 is an equilibrium of (2.1) with $f = f_1$, it follows from Theorem 7.6.1 that the principal eigenvalue of $\lambda w_1 = d\frac{d^2 w_1}{dx^2} - \frac{dw_1}{dx} + f_1(1 - U_1)w_1$, with the usual boundary conditions, is $\lambda = 0$ and $w_1 = U_1 > 0$ is the principal eigenfunction. Since $f_1(1 - U_1) > [f_1(1 - U_1) - U_1 f_1'(1 - U_1)]$, a standard comparison result for eigenvalues (see e.g. Protter and Weinberger (1967), Theorem 15, pg. 38) implies that the scalar eigenvalue problem obtained by setting $w_2 = 0$ in (3.3) has a negative principal eigenvalue.

Arguing as in the previous paragraph, using $f_2(1) > f_2(1 - U_1)$ it follows that $\lambda_0 + f_2(1) > \Lambda_1$. Consequently, if $\Lambda_1 > 0$ then so is $\lambda_0 + f_2(1) > 0$ and therefore E_2 exists.

Suppose hereafter that $\Lambda_1 > 0$. The proof of the remaining assertions breaks up into several parts. As noted before the statement of the theorem, we aim to use Theorem 2.5.3. This requires that we construct an arc of super-equilibria Γ as in (U_-) of section 5 of Chapter 2. Γ will be a line segment through E_1 in the direction of the principal eigenvector corresponding to (3.3). Arguments similar to those of Remark 6.2 of Chapter 7 will be used for this construction. We begin by showing that the principal eigenvector $w = (P_1, P_2)$ of (3.3) can be chosen to satisfy $w \ll_K 0$.

Corresponding to $\lambda = \Lambda_1$, (3.4) has a principal eigenfunction $w = P_2 \gg 0$. We will show that corresponding to $\lambda = \Lambda_1$, (3.3) has an eigenfunction $w = (P_1, P_2)$ where $P_1 \ll 0$. For simplicity of notation, let $a(x) = f_1(1 - U_1(x))$ and $b(x) =$

$U_1(x)f_1'(1-U_1(x))$ and note that $a \gg 0$ and $b \gg 0$. Choose a positive number s such that $s - a(x) > 0$ and $s - b(x) - \Lambda_1 > 0$. Then, the first of equations (3.3), with $\lambda = \Lambda_1$ and $w_2 = P_2$, may be rewritten as

$$-bP_2 = Lw_1 - (s - b - \Lambda_1)w_1 \tag{3.6}$$

where

$$Lw_1 = -\frac{d^2w_1}{dx^2} + \frac{dw_1}{dx} + (s-a)w_1.$$

Denote by $\bar{\lambda}(m)$, the smallest eigenvalue of

$$Lw = \lambda m(x)w$$

together with the usual boundary conditions, where $m(x) > 0$ and continuous. By Theorem 4.5 of Amann(1976),

$$\bar{\lambda}(s - b - \Lambda_1) > \bar{\lambda}(s).$$

Now, $U_1 \gg 0$ satisfies the boundary conditions and, because it is an equilibrium solution of (2.1), it satisfies $LU_1 = 1sU_1$ so it follows from Theorem 7.6.1 that U_1 is the principal eigenfunction and $\bar{\lambda}(s) = 1$. Consequently, from the above inequality, $\bar{\lambda}(s - b - \Lambda_1) > 1$. Then, by Theorem 4.4 of Amann(1976), (3.6) together with boundary conditions, can be solved uniquely for $w_1 \equiv P_1$ and this solution satisfies $P_1 \ll 0$ because $-bP_2 \ll 0$. Thus, $w = (P_1, P_2)$ satisfies (3.3) and the boundary conditions corresponding to $\lambda = \Lambda_1$ and $w \ll_K 0$.

Now we apply Corollary 7.3.6, using the order relation $\leq_K$, to show the existence of a monotone line segment of super-equilibria through E_1 in the direction w, verifying that (U_-) of Proposition 2.5.2 holds. Let $(u_1, u_2) = (U_1, 0) + r(P_1, P_2) = E_1 + rw$. Then, using (2.1) and (3.3), we have

$$d\frac{d^2u_1}{dx^2} - \frac{du_1}{dx} + u_1f_1(1 - u_1 - u_2) = [d\frac{d^2U_1}{dx^2} - \frac{dU_1}{dx} + U_1f_1(1 - U_1)]$$
$$+ r[\frac{d^2P_1}{dx^2} - \frac{dP_1}{dx} + (f_1(1-U_1) - U_1f_1'(1-U_1))P_1 - U_1f_1'(1-U_1)P_2]$$
$$+ (U_1 + rP_2)f_1(1 - U_1 - r(P_1 + P_2)) - U_1f_1(1 - U_1)$$
$$- rP_1[f_1(1-U_1) - U_1f_1'(1-U_1)] + rU_1f_1'(1-U_1)P_2$$
$$= r\Lambda_1P_1 + U_1[f_1(1 - U_1 - r(P_1 + P_2)) - f_1(1 - U_1) + r(P_1 + P_2)f_1'(1-U_1)]$$
$$+ rP_1[f_1(1 - U_1 - r(P_1 + P_2)) - f_1(1-U_1)]$$
$$= r[\Lambda_1P_1 + U_1O(r) + O(r)] < 0$$

and

$$\frac{d^2u_2}{dx^2} - \frac{du_2}{dx} + u_2f_2(1 - u_1 - u_2) = r[(\frac{d^2P_2}{dx^2} - \frac{dP_2}{dx} + P_2f_2(1 - U_1))$$
$$+ P_2(f_2(1 - U_1 - r(P_1 + P_2)) - f_2(1 - U_1))$$
$$= rP_2(\Lambda_1 + O(r)) > 0,$$

where $O(r)$ represents a term satisfying $O(r) \to 0$ as $r \to 0$, uniformly in x. Therefore, the final inequality of each differential inequality holds if r is sufficiently small, say, for $0 < r \leq r_0$. It follows from Corollary 7.3.6 that for each $r \in (0, r_0]$,

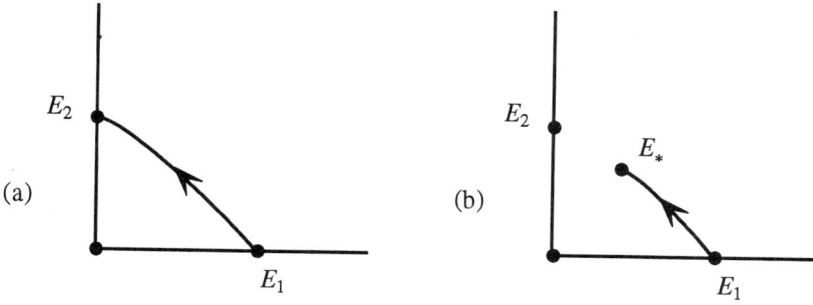

FIGURE 3.1. If $\Lambda_1 > 0$ there is a monotone full orbit connecting E_1 to (a) E_2 or (b) a positive equilibrium E_*.

$E_1 + rw$ is a strict super-solution. Thus, (U_-) of Proposition 2.5.2 holds with $\Gamma = \{E_1 + rw : 0 \leq r \leq r_0\}$.

The other hypotheses of Theorem 2.5.3 must be verified. Equations (1.8)-(1.10) generate a semiflow Φ on X_Ω which is strictly monotone by Corollary 7.3.5. For the following, we are only interested in the restriction of Φ to the positively invariant set $Z = \{u \in X_\Omega : E_2 \leq_K u \leq_K E_1\}$. By Proposition 1.3, Φ is strongly monotone on $\{u \in X_\Omega : E_2 \ll_K u \ll_K E_1\}$. By Proposition 2.5.2 and the remarks following it, there exists an equilibrium E_* in Z, possibly coinciding with E_2, satisfying $E_2 \leq_K E_* \ll_K E_1$ and such that for each $u \in \Gamma$, $u \neq E_1$, we have $\Phi_t(u) \to E_*$ as $t \to \infty$. In fact, $\Phi_{t_2}(u) \ll_K \Phi_{t_1}(u) \ll_K u$ if $t_2 > t_1 > 0$. Furthermore, E_* attracts all solutions corresponding to initial data satisfying (3.5). Indeed, if u satisfies (3.5), then by Proposition 1.3, $u(t) = \Phi_t(u)$ satisfies $E_* \leq u(1) \ll_K E_1$ so $u(1)$ is attracted to E_* by Proposition 2.5.2.

In order to apply Theorem 2.5.3, note first that $\Gamma \subset Z$ and $O(\Gamma) \subset Z$. By Theorem 7.3.1, $\Phi_1(Z)$ has compact closure in Z and therefore, hypothesis (1) of Theorem 2.5.3 holds. We now check that hypotheses (2)-(4), appropriately modified according to the remarks following Theorem 2.5.3, hold. Z is a subset of the Banach space $Y = X = C([0,1], \mathbb{R}^2)$ with cone $Y_+ = \{u = (u_1, u_2) \in Y : u_1 \geq 0, u_2 \leq 0\}$. Hypotheses (2)-(4) deal with the semiflow Φ restricted to $L = (E_1 - \partial Y_+) \cap Z$. A point $u = (u_1, u_2)$ in L satisfies $u_1 \leq U_1$, $u_2 \geq 0$ and either $u_1(x) = U_1(x)$ for some x or $u_2(x) = 0$ for some x. If $u_i > 0$, then $u(t) \ll_K E_1$ for $t > 0$ by Proposition 1.3 where $u(t) = (u_1(t), u_2(t)) = \Phi_t(u)$. If $u_1 \neq U_1$ and $u_2 = 0$, then $u(t) = (u_1(t), 0)$ and $u_1(t) \ll U_1$ for $t > 0$ holds by virtue of the strong monotonicity of the semiflow generated by (2.1)-(2.3). It follows that if $u \in L$, $u \neq E_1$, then either $u(t) = (u_1(t), 0)$ for $t > 0$, where $u_1(t) \ll U_1$, or $u(t) \ll_K E_1$ for all $t > 0$. Since all solutions of the first kind approach E_1 as $t \to \infty$, unless $u = E_0$, by Theorem 2.1, we see that (2) and (3) hold. If there were a full orbit ϕ in $L \cup E_{--}$ satisfying $\phi(t) \to E_1$ as $t \to -\infty$, then necessarily $\phi(t) = (u_1(t), 0)$ and $u_1(t) < u_1(s) < E_1$ if $t, s \in \mathbb{R}$ satisfy $s < t$. The existence of such a solution would contradict the second assertion of Theorem 2.1. Therefore, hypothesis (4) holds. The existence of the full orbit having the properties described in the theorem follows from Theorem 5.2.3. □

Figure 3.1 describes pictorially the assertions of Theorem 3.2.

A symmetrical analysis can be applied to determine the stability properties of the equilibrium E_2 provided $\lambda_0 + f_2(1) > 0$. We describe the essential features informally below. The eigenvalue problem associated with the variational equation about $E_2 = (0, U_2)$ is given by

$$\lambda w_1 = d\frac{d^2 w_1}{dx^2} - \frac{dw_1}{dx} + w_1 f_1(1 - U_2) \qquad (3.7)$$
$$\lambda w_2 = d\frac{d^2 w_2}{dx^2} - \frac{dw_2}{dx} - w_1 f_2'(1 - U_2)U_2 + w_2[f_2(1 - U_2) - U_2 f_2'(1 - U_2)].$$

together with the usual boundary conditions. As with (3.3), this eigenvalue problem is decoupled. The eigenvalues of (3.7) are related to those of the scalar eigenvalue problem

$$\lambda w = d\frac{d^2 w}{dx^2} - \frac{dw}{dx} + w f_1(1 - U_2) \qquad (3.8)$$

with the usual boundary condition. Let Λ_2 denote the principal eigenvalue of (3.8) and Q_1 denote the corresponding principal eigenfunction. If $\Lambda_2 < 0$ then E_2 is asymptotically stable. If $\Lambda_2 > 0$, then so is $\lambda_0 + f_1(1) > 0$ and therefore E_1 exists. In this case, corresponding to the principle eigenvalue Λ_2 of (3.8), there exists a function $Q_2 \ll 0$ such that $w = (Q_1, Q_2)$ satisfies (3.7) corresponding to $\lambda = \Lambda_2$. Furthermore, there exists a full orbit $\phi : \mathbb{R} \to X_\Omega$ of Φ satisfying:
(i) $E_2 \ll_K \phi(t_1) \ll_K \phi(t_2) \ll_K E_1$ if $t_1 < t_2$.
(ii) $\phi(t) \to E_2$, $t \to -\infty$.
(iii) There exists an equilibrium E_{**} of (1.8),(1.9), possibly coinciding with E_1, satisfying $E_2 \ll_K E_{**} \leq_K E_1$ and

$$\phi(t) \to E_{**}, \quad t \to +\infty.$$

Furthermore, E_{**} attracts all solutions of (1.8)-(1.10) for which

$$E_2 \leq_K (u_{10}, u_{20}) \leq_K E_{**} \quad \text{and } u_{10} > 0. \qquad (3.9)$$

4. Coexistence

From a biological stand point, the main question is whether both populations can coexist in the tubular reactor indefinitely. The principal result of this chapter, stated below, says that the answer is yes under suitable conditions. In biological terms these conditions are:
(1) Each population can survive in the culture vessel in the absence of a competitor ($\lambda_0 + f_i(1) > 0$, $i = 1, 2$), and
(2) For each $i \neq j$, population u_i can successfully invade the reactor in which only population u_j is present at equilibrium E_j, if u_i is introduced in infinitesimally small concentration ($\Lambda_j > 0$, $j = 1, 2$).

If (1) and (2) hold, then most solutions converge to a positive equilibrium. Mathematically, the result is stated as follows.

THEOREM 4.1. *Suppose that $\lambda_0 + f_i(1) > 0$ and $\Lambda_i > 0$ for $i = 1, 2$. Then E_i exists for $i = 1, 2$ and there exist positive equilibria E_* and E_{**}, possibly identical, satisfying*

$$E_2 \ll_K E_{**} \leq_K E_* \ll_K E_1.$$

E_* attracts all solutions of (1.8)-(1.10) corresponding to initial data satisfying (3.5) and E_{**} attracts all solutions corresponding to initial data satisfying (3.9). The omega limit set of every solution for which $u_{i0} > 0$ for $i = 1, 2$ is contained in the set
$$O = \{u_0 \in X_\Theta : E_{**} \leq_K u_0 \leq_K E_*\}.$$
Furthermore, there is an open and dense subset of X_Θ consisting of initial data for which the corresponding solution converges to an equilibrium belonging to O. In case $E_* = E_{**}$, then E_* attracts all solutions of (1.8)-(1.10) for which the corresponding initial data satisfy $u_{i0} > 0$ for $i = 1, 2$.

PROOF. We begin by identifying those assertions that follow immediately from Theorem 3.2 and the symmetrical analysis applied to E_2 described following that theorem. Since both $\Lambda_1 > 0$ and $\Lambda_2 > 0$, both these analyses apply and we use the notation P and Q for the corresponding principal eigenvectors. For small positive r, we have $E_2 \ll_K E_2 + rQ \ll_K E_1 + rP \ll_K E_1$ and therefore the monotonicity of Φ implies that $E_2 \ll_K \Phi_t(E_2 + rQ) \ll_K \Phi_t(E_1 + rP) \ll_K E_1$ for $t > 0$. Since $\Phi_t(E_2+rQ) \to E_{**}$ and $\Phi_t(E_1+rP) \to E_*$ as $t \to \infty$ monotonically, it follows that $E_2 \ll_K E_{**} \leq_K E_* \ll_K E_1$, where equality $E_{**} = E_*$ may hold. The assertions concerning the domains of attraction of E_{**} and E_* follow also from Theorem 3.2 and the remarks following it.

If $u_0 = (u_{10}, u_{20}) \in X_\Theta$, then $(0, u_{20}) \leq_K u_0 \leq_K (u_{10}, 0)$ and by the monotonicity of Φ (Proposition 1.3) it follows that $\Phi_t((0, u_{20})) \leq_K \Phi_t(u_0) \leq_K \Phi_t((u_{10}, 0))$ for $t \geq 0$. But $\Phi_t((0, u_{20})) \to E_2$ as $t \to \infty$, by Theorem 2.1, unless $u_{20} = 0$, in which case, $\Phi_t((0, u_{20})) \equiv 0$. Similarly, $\Phi_t((u_{10}, 0)) \to E_1$ as $t \to \infty$, unless $u_{10} = 0$. As the solution $\Phi_t(u_0)$ is bounded for $t \geq 0$, its omega limit set is nonempty, compact and invariant. By the inequalities above, it must be the case that $E_2 \leq_K v \leq_K E_1$ for all $v \in \omega(u_0)$.

Now we observe that if u_0 is such that $u_{i0} \neq 0$ for $i = 1, 2$ and $E_2 \leq_K u_0 \leq_K E_1$, then $\omega(u_0) \subset O$. In fact, if $u(t) = (u_1(t), u_2(t))$ is the corresponding solution of (1.8)-(1.10), then $0 \ll u_i(t)$ for $t > 0$ by Proposition 1.2. But then, by the conditional strong monotonicity (Proposition 1.3) and the fact that $u(1)$ is positive and $E_2 <_K u(1) <_K E_1$, we conclude that $E_2 \ll_K u(2) \ll_K E_1$. Consequently, by choosing $r > 0$ sufficiently small we have $E_2 + rQ \leq_K u(2) \leq_K E_1 + rP$. Monotonicity of Φ implies that $\Phi_t(E_2 + rQ) \leq_K u(t+2) \leq_K \Phi_t(E_1 + rP)$ for $t \geq 0$ and therefore, $E_{**} \leq_K v \leq_K E_*$ for every $v \in \omega(u_0)$ since $\Phi_t(E_2 + rQ) \to E_{**}$ and $\Phi_t(E_1 + rP) \to E_*$ as $t \to \infty$.

We claim that $\omega(u_0) \subset O$ if it contains any point $v = (v_1, v_2)$ satisfying $v_i \neq 0$ for $i = 1, 2$. Indeed, since the omega limit set is invariant, the solution starting at v belongs to the omega limit set so by Proposition 1.2 we may as well assume that $0 \ll v_i$ for $i = 1, 2$. $E_2 \leq_K v \leq_K E_1$, from the first paragragh of the proof, and therefore, by the argument of the previous paragragh we conclude that $\omega(v) \subset O$. But $\omega(v) \subset \omega(u_0)$ and therefore the solution $u(t)$ of (1.8)-(1.10) with $u(0) = u_0$ gets arbitrarily close to some point of O. Consequently, there exists $r > 0$ and $t_0 > 0$ such that $E_2 + rQ \ll_K u(t_0) \ll_K E_1 + rP$ since $\{w \in X_\Theta : E_2 + rQ \ll_K w \ll_K E_1 + rP\}$ is an open neighborhood of O in X_Θ. Therefore, by the previous paragragh, $\omega(u_0) = \omega(u(t_0)) \subset O$.

Thus far, we have established that either $\omega(u_0) \subset O$ or $\omega(u_0) \subset X_1 \cup X_2$, where $X_i = \{u = (u_1, u_2) \in X_\Theta : u_j = 0\}$ and $i \neq j$. $E_0 \notin \omega(u_0)$ unless $u_0 = 0$ because E_0 is a repellor by Theorem 3.1. Therefore, as the omega limit set is connected, if

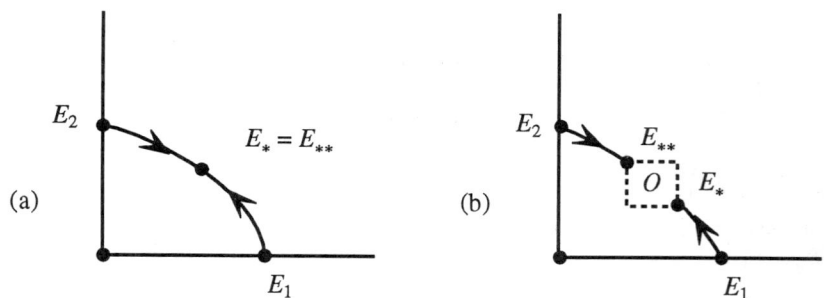

FIGURE 4.1. An illustration of Theorem 4.1 when there is a unique positive equilibrium (a) and when there are more than one positive equilibria (b).

$u_0 \neq 0$ then either $\omega(u_0) \subset X_1 \setminus E_0$ or $\omega(u_0) \subset X_2 \setminus E_0$. An omega limit set is also invariant and by Proposition 2.1, the only invariant subset of $X_i \setminus E_0$ is E_i. Hence, if $u_0 \neq 0$, then either $\omega(u_0) = O$ or $\omega(u_0) = E_i$ for some i.

We now seek to exclude the possibility that $\omega(u_0) = E_i$ in the case that $u_0 = (u_{10}, u_{20})$ where $u_{i0} \neq 0$. Let $u(t) = (u_1(t), u_2(t))$ be the corresponding solution and assume that $u(t) \to E_1$ as $t \to \infty$. The other case is handled similarly. Fix $s > 0$ so small that the largest eigenvalue, denoted by c, of

$$\lambda w = d\frac{d^2 w}{dx^2} - \frac{dw}{dx} + f_2(1 - U_1 - s)w,$$

together with the usual boundary conditions, is positive. Here, we use the continuity of the principal eigenvalue with respect to the "potential" and the fact that $\Lambda_1 > 0$. Let w be the corresponding principal eigenfunction satisfying $0 \ll w$. Now, there exists $t_0 > 0$ such that for all $t \geq t_0$, we have $f_2(1 - u_1 - u_2) \geq f_2(1 - U_1 - s)$ since $u_1(t) + u_2(t) \to U_1$ as $t \to \infty$. Let $\delta > 0$ be such that $u_2(t_0) \geq \delta w$. That such a δ can be found follows from the fact that $u_2(t) \gg 0$ for $t > 0$. Then, for $t \geq t_0$, u_2 satisfies

$$\frac{\partial u_2}{\partial t} \geq d\frac{\partial^2 u_2}{\partial x^2} - \frac{\partial u_2}{\partial x} + u_2 f_2(1 - U_1 - s),$$

the usual boundary conditions, and $v(t_0) \geq \delta w$. By Theorem 7.3.4, we may conclude that $u_2(t) \geq \delta w \exp(c(t - t_0))$ for $t \geq t_0$. But this implies that $u_2(t)$ is unbounded and this contradiction proves that $\omega(u_0) \neq E_i$. Thus, $\omega(u_0) \subset O$ if $u_{i0} \neq 0$, $i = 1, 2$.

Finally, the assertion that there is an open and dense subset of X_Θ consisting of convergent points follows from Theorem 2.4.7 and Remark 4.1 of Chapter 2. In fact, these results are applied to Φ restricted to the space $X_\Theta \setminus (X_1 \cup X_2)$ on which it is strongly monotone. The arguments establishing that the hypotheses of these results hold are as in the proof of Theorem 7.5.1. □

Figure 4.1 describes pictorially the content of Theorem 4.1.

It is important to point out that it is a nontrivial task to verify that the hypotheses, $\Lambda_i > 0$ for $i = 1, 2$, of Theorem 4.1 are met in any given situation. The numerical simulations of Kung and Baltzis(1992) suggest that they can be met. See that paper for functions and parameter regions for which there are apparently unique stable positive equilibrium solutions. A further point bears mentioning. If the hypotheses of Theorem 4.1 are met for a given choice of parameters and functions f_i, then they are also met if these parameters and functions are perturbed slightly. More precisely, if the functions $g_i(S)$ are sufficiently close to the functions $f_i(S)$ in the C^1 topology on the interval $0 \leq S \leq 1$ and the perturbation in the diffusion coefficient is sufficiently small, then the system (1.8)-(1.10) with g_i in place of f_i and the new diffusion coefficient also satisfies the hypotheses of Theorem 4.1. This is due to the continuity of the principal eigenvalue to perturbation of the coefficients. Thus the conditions of the Theorem are stable under perturbation.

Theorem 4.1 implies the existence of at least one positive equilibrium. Using the arguments of the last paragragh of the proof of Theorem 2.1, one can show that, at a positive equilibrium, both population densities are strictly increasing in x on $0 \leq x \leq 1$, reaching maximum density at $x = 1$. This implies that the nutrient concentration is strictly decreasing on that interval. The question of the uniqueness of a positive equilibrium is unresolved.

5. Remarks and Discussion

As mentioned in the introduction, the model was formulated by Kung and Baltzis (1992), who employed numerical simulations to investigate the dynamics. Actually it is a special case of a slightly more general model considered there. The more general model accounts for the fact that some of the energy derived from the nutrient which a microorganism consumes must go toward its metabolism and maintenance and therefore, is unavailable for cell division. Mathematically, this means that in the equation for u_i, the term $u_i f_i(S)$ becomes $u_i(f_i(S) - \lambda_i)$. The inclusion of this term renders our analysis inapplicable since it is based on the conservation principle: $S(t) + u_1(t) + u_2(t) \to S^0$ as $t \to \infty$. Similarly, our analysis does not apply if the diffusion coefficients of nutrient and organisms differ significantly or if chemotaxis plays a significant role.

The model of Kung and Baltzis is closely related to another model proposed by So and Waltman(1989) and later analyzed by Hsu and Waltman(1993) and Hsu et al(1995). The treatment in this chapter follows the latter paper.

The techniques used in the proof of Theorem 4.1 are widely known. Results of a similar nature are proved by Cantrell and Cosner(1993), Hess(1991), Smith et al(1991), Takeuchi and Lu(1994), Zhou and Pao(1982), to name only a few.

Theorem 4.1 describes the behavior of solutions of (1.8)-(1.10) when $\Lambda_i > 0$ for $i = 1, 2$. There are three other possibilities that may occur, ignoring equality, namely $\Lambda_1 > 0 > \Lambda_2$; $\Lambda_1 < 0 < \Lambda_2$; and $\Lambda_i < 0, i = 1, 2$. Figure 8.2 indicates one possibility when $\Lambda_1 > 0 > \Lambda_2$ holds. Theorem 3.2 applies in this case. If $\Lambda_i < 0$ for $i = 1, 2$ then there must exist an unstable positive equilibrium. See Chapter 6 of Smith and Waltman(1995) for further results in this direction for a related model system.

An abstract approach to competition is taken in Hess and Lazer(1991). In this paper some of the results obtained in this chapter are shown to hold for a general infinite dimensional discrete model of competition. Furthermore, all the cases mentioned in the previous paragraph are treated. In particular, the existence of an unstable equilibrium is established when $\Lambda_i < 0$ for $i = 1, 2$. An important open problem is to give sufficient conditions for the uniqueness of a positive equilibrium when $\Lambda_i > 0$ for $i = 1, 2$ in the generality considered by Hess and Lazer(1991).

It is not immediately clear that the behavior of solutions (1.4)-(1.6) is determined by the behavior of solutions of the limiting system (1.8)-(1.10). That this is so under suitable conditions is justified in Mischaikow et al(1994) and Thieme(1992).

Appendix. Chain Recurrence

In this appendix it is proved that a compact limit set of a flow ψ_t on a metric space X is chain recurrent. The result was originally proved by Conley (see Conley(1972) and Conley(1978)). The proof below, taken from Mischaikow et al (1994), follows one given in Robinson(1977) for mappings. As this result is needed in section 3 of Chapter 3, we use the notation from that section. We begin by recalling the definition of chain recurrence from section 3 of Chapter 3.

Let A be a nonempty invariant subset of X and $x, y \in A$. For $\epsilon > 0$, $t > 0$, an (ϵ, t)-chain from x to y in A is a sequence $\{x = x_1, x_2, \ldots, x_{n+1} = y; t_1, t_2, \ldots, t_n\}$ of points $x_i \in A$ and times $t_i \geq t$ such that $d(\psi_{t_i}(x_i), x_{i+1}) < \epsilon$, $i = 1, 2, \ldots, n$. A point $x \in A$ is called a *chain-recurrent point* if for every $\epsilon > 0$, $t > 0$ there is an (ϵ, t)-chain from x to x in A. The set A is said to be chain recurrent if every point in A is chain-recurrent in A.

The main result of this appendix follows.

THEOREM 1. *Let L be an (omega or alpha) limit set of a (positive or negative) orbit of the flow ψ that has compact closure in X. Then L is chain recurrent.*

Our proof uses the following intuitive result.

LEMMA. *Assume that $\gamma^+(x)$ has compact closure in X and let $T > 0$ and $\gamma_T^+(x) = \{\psi_t(x) : t \geq T\}$. Given $y \in \omega(x)$ and $\epsilon > 0$, $t_0 > 0$, there exists an (ϵ, t_0)-chain*

$$\{y = y_1, y_2, \ldots, y_l, y_{l+1} = y; t_1, t_2, \ldots, t_l\}$$

such that $y_i \in \gamma_T^+(x)$ for $i = 2, \ldots, l$, $t_i = t_0$, $1 \leq i \leq l-1$, $t_0 \leq t_l < 2t_0$.

PROOF. Suppose that $y = \lim_{n\to\infty} \psi_{s_n}(x)$ where $s_n \to \infty$, $n \to \infty$. Choose n such that $s_n > T$ and $d(\psi_{s_n+t}(x), \psi_t(y)) < \epsilon$ for $0 \leq t \leq t_0$. Set $y_1 = y$ and $y_2 = \psi_{s_n+t_0}(x)$ and $t_1 = t_0$. Then

$$d(\psi_{t_1}(y_1), y_2) = d(\psi_{t_0}(y), \psi_{s_n+t_0}(x)) < \epsilon.$$

Choose m such that $s_m > s_n + 2t_0$ and $d(\psi_{s_m}(x), y) < \epsilon$. Let $k \geq 1$ be such that $s_m - s_n - t_0 = kt_0 + r$ for some r, $0 \leq r < t_0$. Let $y_3 = \psi_{s_n+2t_0}(x)$, $y_4 = \psi_{s_n+3t_0}(x), \ldots, y_{k+1} = \psi_{s_n+kt_0}(x)$, $y_{k+2} = y$, and $t_i = t_0$, $1 \leq i \leq k$, $t_{k+1} = t_0 + r$. Then $d(\psi_{t_i}(y_i), y_{i+1}) = 0$, $i = 2, \ldots, k$ and $d(\psi_{t_{k+1}}(y_{k+1}), y_{k+2}) = d(\psi_{s_m}(x), y) < \epsilon$. □

PROOF OF THEOREM 1. We begin by treating the case that $L = \omega(x)$ where $\gamma^+(x)$ has compact closure in X. The other case can be reduced to this one by time reversal, as will be shown in the last paragragh of the proof.

163

Let $y \in \omega(x)$ and $\epsilon > 0$, $t_0 > 0$. By the Lemma, for each $n = 1, 2, \ldots$, there exists a $(\frac{1}{n}, t_0)$-chain belonging to $\gamma_n^+(x)$ having the properties described in the Lemma with $T = n$. Let y_i^n, $1 \leq i \leq l_n + 1$, be the points of the $(\frac{1}{n}, t_0)$-chain, t_i^n be the times, $1 \leq i \leq l_n$, ($t_i^n = t_0, 1 \leq i \leq l_n - 1, t_0 \leq t_{l_n}^n < 2t_0$) and set $C^n = \{y_i^n : 1 \leq i \leq l_n\}$.

Since $C^n \subset \overline{\gamma^+(x)}$ and $\overline{\gamma^+(x)}$ is compact in X, by passing to a subsequence, if necessary, we can assume that $C^n \to C$ as $n \to \infty$ in the Hausdorff metric on the space of closed subsets of $\overline{\gamma^+(x)}$, where C is a nonempty compact subset of $\overline{\gamma^+(x)}$. In fact, as $y \in C^n \subset \overline{\gamma_n^+(x)}$ it follows that $y \in C \subset \omega(x)$.

Choose δ, $0 < \delta < \epsilon/3$, such that whenever $z_1, z_2 \in \overline{\gamma^+(x)}$ and $d(z_1, z_2) < \delta$ then $d(\psi_t(z_1), \psi_t(z_2)) < \epsilon/3$ for $0 \leq t \leq 2t_0$. Fix n such that $1/n < \epsilon/3$ and $D(C^n, C) < \delta$, where D denotes the Hausdorff metric. Drop the superscript n on y_i^n and t_i^n. Then we have

$$d(\psi_{t_1}(y_1), y_2) = d(\psi_{t_0}(y), y_2) < \frac{1}{n} < \frac{\epsilon}{3}.$$

Set $z_1 = y$ and choose $z_2 \in C$ such that $d(z_2, y_2) < \delta < \epsilon/3$. Then

$$d(\psi_{t_1}(z_1), z_2) = d(\psi_{t_0}(y), z_2)$$
$$\leq d(\psi_{t_0}(y), y_2) + d(y_2, z_2)$$
$$< 2\epsilon/3.$$

Since $d(z_2, y_2) < \delta$, it follows that

$$d(\psi_{t_2}(z_2), \psi_{t_2}(y_2)) < \epsilon/3.$$

Choose $z_3 \in C$ such that $d(z_3, y_3) < \delta < \epsilon/3$. Then

$$d(\psi_{t_2}(z_2), z_3) \leq d(\psi_{t_2}(z_2), \psi_{t_2}(y_2))$$
$$+ d(\psi_{t_2}(y_2), y_3)$$
$$+ d(y_3, z_3)$$
$$< \frac{\epsilon}{3} + \frac{1}{n} + \delta < \epsilon.$$

Furthermore, $d(z_3, y_3) < \delta$ implies

$$d(\psi_{t_3}(z_3), \psi_{t_3}(y_3)) < \epsilon/3,$$

so we choose $z_4 \in C$ such that $d(z_4, y_4) < \delta < \epsilon/3$. As above, $d(\psi_{t_3}(z_3), z_4) < \epsilon$. Clearly, we may continue in this way, finding $z_i \in C$ with $d(z_i, y_i) < \delta$, $i = 1, 2, \ldots, l = l_n$ and t_i, $i = 1, 2, \ldots, l - 1$, such that $d(\psi_{t_i}(z_i), z_{i+1}) < \epsilon$ for $i = 1, 2, \ldots, l - 1$. Since $d(z_l, y_l) < \delta$, we may conclude $d(\psi_{t_l}(z_l), \psi_{t_l}(y_l)) < \epsilon/3$. Set $z_{l+1} = y$ and observe that

$$d(\psi_{t_l}(z_l), z_{l+1}) \leq d(\psi_{t_l}(z_l), \psi_{t_l}(y_l))$$
$$+ d(\psi_{t_l}(y_l), y)$$
$$< \frac{\epsilon}{3} + \frac{1}{n} < \frac{2}{3}\epsilon.$$

We have constructed an (ϵ, t_0)-chain in $C \subset \omega(x)$ joining the point y to itself. Since $y \in \omega(x)$, $\epsilon > 0$, $t_0 > 0$ were arbitrary, it follows that $\omega(x)$ is chain recurrent.

The following ideas show that the case that $L = \alpha(x)$ can be reduced to the former case. First note that if $L = \alpha(x)$ for the flow ψ_t then $L = \omega(x)$ for the flow $\phi_t \equiv \psi_{-t}$ and therefore L is chain recurrent for the flow ϕ_t by the proof given above. We must show that L is chain recurrent for ψ_t. Fix $y \in L$, $\epsilon > 0$ and $t > 0$. We will show that there is an (ϵ, t)-chain from y to y. Since L is compact, we can choose $\delta > 0$ such that $\delta < \epsilon$ and whenever $x_1, x_2 \in L$ and $d(x_1, x_2) < \delta$ then $d(\psi_t(x_1), \psi_t(x_2)) < \epsilon$. Now, since y is a chain recurrent point for ϕ_t, there exists a $(\delta, 2t)$-chain $\{y = x_1, x_2, \ldots, x_{n+1} = y;\ t_1, t_2, \ldots, t_n\}$ of points $x_i \in L$ and times $t_i \geq 2t$ such that $d(\psi_{-t_i}(x_i), x_{i+1}) < \delta$, $i = 1, 2, \ldots, n$. Let $y_2 = \psi_{-t_n + t}(x_n)$, $y_i = \psi_{-t_{n+2-i}}(x_{n+2-i})$ for $i = 3, 4 \ldots, n+1$, $y_1 = y_{n+2} = y$, $s_1 = t$, $s_2 = t_n - t$, $s_i = t_{n+2-i}$ for $3 \leq i \leq n+1$. Then, $s_i \geq t$ for $1 \leq i \leq n+1$ and $d(\psi_{s_1}(y_1), y_2) = d(\psi_t(y), \psi_t(\psi_{-t_n}(x_n))) < \epsilon$ because $d(y, \psi_{-t_n}(x_n)) < \delta$. Also, $d(\psi_{s_2}(y_2), y_3) = d(x_n, \psi_{-t_{n-1}}(x_{n-1})) < \delta < \epsilon$. For $3 \leq i \leq n$, $d(\psi_{s_i}(y_i), y_{i+1}) = d(x_{n+2-i}, \psi_{-t_{n-i+1}}(x_{n-i+1})) < \delta < \epsilon$. Finally, $d(\psi_{s_{n+1}}(y_{n+1}), y_{n+2}) = d(x_1, y) = 0$. Thus, $\{y_1, y_2, \ldots, y_{n+2}; s_1, s_2, \ldots, s_{n+1}\}$ is an (ϵ, t)-chain from y to y for the flow ψ. It follows that y is a chain recurrent point for the flow ψ. $\square$

Bibliography

F. Albrecht, H. Gatzke, A. Haddad, and N. Wax, *The dynamics of two interacting populations*, J. Math. Anal. Appl. **46** (1974), 658–670.

H. Amann, *Fixed Point equations and nonlinear eigenvalue problems in ordered Banach Spaces*, SIAM Rev. **18** (1976), 620–709.

_____, *Invariant sets and existence theorems for semilinear parabolic and elliptic systems*, J. Math. Anal. Appl. **65** (1978), 432–467.

G. Aronsson and R. B. Kellogg, *On a differential equation arising from compartmental analysis*, Math. Biosciences **38** (1978), 1131-122.

A. Arneodo, P. Coullet, J. Peyraud, and C. Tresser, *Strange attractor in Volterra Equations for species in competition*, J. Math. Biol. **14** (1982), 153–157.

C. Batty and D. Robinson, *Positive one-parameter semigroups on ordered spaces*, Acta Appl. Math. **2** (1984), 221–296.

A. Berman and R. Plemmons, *Nonnegative matrices in the mathematical sciences*, Academic Press, New York, 1979.

P. Brown, *Decay to uniform states in ecological interactions*, SIAM J. Appl. Math. **38** (1980), 22–37.

G. Butler and P. Waltman, *Persistence in dynamical systems*, J. Diff. Eqn. **63** (1986), 255–263.

Ph. Clements, H. Heijmans, S. Angenent, C. van Duijn, and B. de Pagter, *One-Parameter Semigroups*, CWI Monographs 5, North-Holland, Amsterdam, 1987.

E. Coddington and N. Levinson, *Theory of Ordinary Differential Equations*, McGraw Hill, New York, 1955.

C. Conley, *The gradient structure of a flow :I*, Also, Ergodic Theory and Dynamical Systems, 8, 1988, 11-26. (1972), IBM Research, RC 3939 (17806) Yorktown Heights, NY.

_____, *Isolated Invariant Sets and the Morse Index*, CBMS **38** (1978), Amer. Math. Soc., Providence,R.I..

E. Conway and J. Smoller, *A comparison theorem for systems of reaction-diffusion equations*, Comm. Part. Diff. Eqns. **2** (1977), 679–697.

E. Conway, D. Hoff, and J. Smoller, *Large time behavior of solutions of systems of nonlinear reaction-diffusion equations*, SIAM J. Appl. Math. **35** (1978), 1–16.

W. A. Coppel, *Stability and Asymptotic Behavior of Differential Equations*, Heath, Boston, 1965.

R. S. Cantrell and C. Cosner, *Should a park be an island?*, SIAM J. Appl. Math. **53** (1993), 219–252.

E. Dancer and P. Hess, *Stability of fixed points for order-preserving discrete-time dynamical systems*, J. reine angew. Math. **419** (1991), 125–139.

A. Friedman, *Partial Differential Equations of Parabolic Type*, Prentice Hall, Englewood Cliffs, N.J., 1964.

_____, *Remarks on nonlinear parabolic equations*, Proc. of Symposia in Applied Math., XVII, Application of Nonlinear Partial Differential Equations in Mathematical Physics, Amer. Math. Soc., 1965.

P. Fife, *Mathematical aspects of reacting and diffusing systems*, Lecture Notes in Math. **28** (1979), Springer-Verlag, New York.

P. Fife and M. Tang, *Comparison principles for reaction-diffusion systems: irregular comparison functions and applications to questions of stability and speed of propagation of disturbances*, 40 **J. Diff. Eqns.** (1981), 168–185.

R. Gardner, *Comparison and stability theorems for reaction-diffusion systems*, SIAM J. Math. Anal. **12** (1980), 60–69.

B. S. Goh, *Stability in models of mutualism*, Amer. Naturalist **113** (1979), 261–275.

G. Greiner, *Zur Perron-Frobenius Theorie stark stetiger Halbgruppen*, Math. Zeit. **177** (1981), 401–423.

S. Grossberg, *Competition, Decision, and Consensus*, J. Math. Anal. Appl. **66** (1978), 470–493.

_____, *Biological Competition: Decision rules, pattern formation, and oscillations*, Proc. Nat. Acad. Sci. **77** (1980), 2338–2342.

K. Hadeler and J. Tomiuk, *Periodic solutions to difference-differential equations*, Arch. Rational Mech. Anal. **65** (1977), 87–95.

K. Hadeler and D. Glas, *Quasimonotone systems and convergence to equilibrium in a population genetics model*, J. Math. Anal. Appl. **95** (1983), 297–303.

N. Hayes, *Roots of transcendental equations associated with certain differential difference equations*, J. London Math. Soc. **25** (1950), 226–232.

W. Gurney, S. Blythe and R. Nisbet, *Nicholson's blowflies revisited*, Nature **287** (1980), London, 17–21.

J. K. Hale, *Ordinary Differential Equations*, Krieger, Malabar, Florida, 1980.

_____, *Asymptotic Behavior of Dissipative Systems publ Amer. Math. Soc.*, Providence, 1988.

_____, *Theory of Functional Differential Equations*, Springer-Verlag, New York, 1977.

P. Hess, *Periodic-parabolic Boundary Value Problems and Positivity*, Longman Scientific and Technical, New York, 1991.

P. Hess and A. C. Lazer, *On an abstract competition model and applications*, Nonlinear Analysis T.M.A. **16** (1991), 917–940.

M. Hirsch, *Systems of differential equations which are competitive or cooperative 1: limit sets*, SIAM J. Appl. Math. **13** 167–179 (1982).

_____, *Systems of differential equations which are competitive or cooperative II: convergence almost everywhere*, SIAM J. Math. Anal. **16** (1985), 423–439.

_____, *The dynamical systems approach to differential equations*, Bull. A.M.S. **11** (1984), 1–64.

_____, *Systems of differential equations which are competitive or cooperative III. Competing species*, Nonlinearity **1** (1988a), 51-71.

_____, *Stability and Convergence in Strongly Monotone dynamical systems*, J. reine angew. Math. **383** (1988b), 1–53.

_____, *Systems of differential equations that are competitive or cooperative. IV: Structural stability in three dimensional systems*, SIAM J. Math. Anal.21 (1990), 1225–1234.

_____, *Systems of differential equations that are competitive or cooperative. V: Convergence in 3-dimensional systems*, J. Diff. Eqns. **80** (1989), 94–106.

_____, *Systems of differential equations that are competitive or cooperative. VI: A local C^r closing lemma for 3-dimensional systems*, Ergod. Th. Dynamical Sys. **11** (1991), 443–454.

M. Hirsch and S. Smale, *Differential Equations, Dynamical Systems, and Linear Algebra*, Acad. Press, New York, 1974.

S.-B. Hsu and P. Waltman, *Analysis of a model of two competitors in a chemostat with an external inhibitor*, SIAM J. Appl. Math. **52** (1992), 528–540.

_____, *On a system of reaction-diffusion equations arising from competition in an unstirred chemostat*, SIAM J. Appl. Math. **53** (1993), 1026–1044.

S.-B. Hsu, H. Smith, and P. Waltman, *Dynamics of competition in the unstirred chemostat*, Canadian Applied Math. Quart. (1994) (to appear).

W. Jager, W.-H. So, B. Tang, and P. Waltman, *Competition in the Chemostat*, J. Math. Biol. **25** (1987), 23–42.

J. Jiang, *Attractors for strongly monotone flows*, J. Math. Anal. Appl. **162** (1991), 210–222.

_____, *On the existence and uniqueness of connecting orbits for cooperative systems*, Acta Mathematica Sinica, New Series **8** (1992), 184–188.

_____, *Three and four dimensional cooperative systems with every equilibrium stable*, J. Math. Anal. Appl. (1994) (to appear).

L. Ladyzhenskaya, *Attractors for semigroups and evolution equations*, Cambridge Univ. Press, Cambridge, 1991.

A. W. Leung, *Systems of nonlinear partial differential equations: applications to biology and engineering* (1989), Kluwer Academic Publishers, Dordrecht, Boston.

Y. M. Li and J. Muldowney, *Global Stability for the SEIR Model in Epidemiology*, J. Math. Biol. (1994) (to appear).

P. L. Lions, *Structure of the set of steady-state solutions and asymptotic behavior of semilinear heat equations*, J. Diff. Eqns. **53** (1984), 362–386.

E. Kamke, *Zur Theorie der Systeme Gewoknlicher Differentialgliechungen, II*, Acta Math. **58** (1932), 57–85.

T. Kato, *Perturbation Theory for Linear Operators*, Springer-Verlag, Berlin, 1976.

W. Kerscher and R. Nagel, *Asymptotic behavior of one-parameter semigroups of positive operators*, Acta Applicandae Math. **2** (1984), 297–309.

K. Kishimoto and H. Weinberger, *The spatial homogeneity of stable equilibria of some reaction-diffusion systems on convex domains*, J. Diff. Eqns. **58** (1985), 15–21.

A. Kolmogorov, *Sulla teoria di Volterra della lotta per l'istenzia*, Giorn. Ist. Ital. Attuari **7** (1936), 74–80.

M. Krasnoselskii, *Positive solutions of operator equations*, Groningen, Noordhoff, 1964.

_____, *The operator of translation along trajectories of differential equations*, vol. 19, Transl. Math. Monographs, Providence, 1968.

M. G. Krein and M. A. Rutman, *Linear Operators leaving invariant a cone in a Banach space*, Uspekhi Mat. Nauk 3,3-95, A.M.S. Translation No.26, Amer. Math. Soc., Providence, R.I., 1950.

Y. Kuang, *Delay Differential Equations with Applications in Population Biology*, Mathematics in Science and Engineering, vol. 191, Academic Press, 1993.

C.-M. Kung and B. Baltzis, *The growth of pure and simple microbial competitors in a moving distributed medium*, Math. Biosci. **111** (1992), 295–313.

K. Kunisch and W. Schappacher, *Order preserving evolution operators of functional differential equations*, Boll. Un. Mat. Ital. **B(6)** (1979), 480–500.

R. H. Martin, *Nonlinear Operators and Differential Equations in Banach Spaces*, Wiley and Sons, New York, 1976.

_____, *Asymptotic Stability and Critical Points for nonlinear quasimonotone parabolic systems*, J. Diff. Eqns. **30** (1978), 391–423.

_____, *A maximum principle for semilinear parabolic systems*, Proc. Amer. Math. Soc. **74** (1979), 66–70.

_____, *Asymptotic behavior of solutions to a class of quasimonotone functional differential equations*, Proc. Workshop on Functional Differential equations and nonlinear semigroups, Research Notes in Math. **48** (1981), Pitman, Boston, Ma..

_____, *Asymptotic behavior of solutions to quasimonotone parabolic systems*, Abstract Cauchy Problems and Functional Differential Equations (F. Kappel and W. Schappacher, eds.), Pitman Research Notes in Mathematics, vol. 48, Boston, 1981, pp. 91–111.

R. H. Martin and H. L. Smith, *Abstract Functional Differential Equations and Reaction-Diffusion Systems*, Trans. Amer. Math. Soc. **321** (1990), 1–44.

R. H. Martin and H. L. Smith, *Reaction-diffusion systems with time-delays: monotonicity, invariance, comparison and convergence*, J. reine angew. Math. **413** (1991a), 1–35.

_____, *Convergence in Lotka-Volterra systems with diffusion and delay*, Proc. of Workshop on Differential Equations and Applications Retzhof, Austria, 1989 (1991b), Marcel Dekker.

H. Matano, *Convergence of solutions of one-dimensional semilinear parabolic equations*, J. Math. Kyoto Univ. **18-2** (1978), 221–227.

_____, *Asymptotic behavior and stability of solutions of semilinear diffusion equations*, Publ. Res. Inst. Math. Sci. **19** (1979), 645–673.

_____, *Existence of nontrivial unstable sets for equilibriums of strongly order preserving systems*, J. Fac. Sci. Univ. Tokyo **30** (1984), 645–673.

_____, *Strongly order-preserving local semi-dynamical systems-Theory and Applications*, Res. Notes in Math., 141, Semigroups, Theory and Applications (H. Brezis, M. G. Crandall and F. Kappel, eds.), vol. 1, Longman Scientific and Technical, London, 1986, pp. 178–185.

_____, *Strong Comparison Principle in Nonlinear Parabolic Equations*, Pitman Research Notes in Mathematics, Nonlinear Parabolic Equations: Qualitative Properties of Solutions (L. Boccardo and A. Tesei, eds.), Longman Scientific and Technical, London, 1987, pp. 148–155.

H. Matano and M. Mimura, *Pattern formation in competition-diffusion systems in nonconvex domains*, Publ. of Research Inst. for Math. Sci., Kyoto Univ. **19** (1983), 1049–1079.

E. J. McShane, *Extension of range of functions*, Bull. Amer. Math. Soc. **40** (1934), 837–842.

J. Mierszyński, *On monotone trajectories*, Proc. Amer. Math. Soc. **113** (1991), 537–544.

K. Mischaikow, H. L. Smith and H. R. Thieme, *Asymptotically autonomous semiflows: Chain recurrence and Lyapunov functions*, Trans. Amer. Math. Soc. (to appear).

X. Mora, *Semilinear Parabolic problems define semiflows on C^k spaces*, Trans. Amer. Math. Soc. **278** (1983), 21-55.

M. Müller, *Uber das fundamenthaltheorem in der theorie der gewohnlichen differentialgleichungen*, Math. Zeit. **26** (1926), 619–645.

J. Murray, *Mathematical Biology*, Springer-Verlag, Berlin, 1989.

R. Nisbet and W. Gurney, *Modelling Fluctuating Populations*, John Wiley and Sons, New York, 1982.

Y. Ohta, *Qualitative analysis of nonlinear quasimonotone dynamical systems described by functional differential equations*, IEEE Trans. Circuits and Systems **28** (1981), 138–144.

C. V. Pao, *Nonlinear parabolic and elliptic equations*, Plenum Press, New York, 1992.

A. Pazy, *Semigroups of Linear Operators and Applications to Partial Differential Equations*, Springer-Verlag, New York, 1983.

P. Poláčik, *Convergence in Smooth Strongly monotone flows defined by semilinear parabolic equations*, J. Diff. Eqns. **79** (1989a), 89–110.

_____, *Domains of attraction of equilibria and monotonicity properties of convergent trajectories in parabolic systems admitting strong comparison principle*, J. reine angew. Math. **400** (1989b), 32–56.

_____, *Existence of unstable sets for invariant sets in compact semiflows. Applications in order-preserving semiflows*, Comment. Math. Univ. Carolinae **31** (1990), 263–276.

M. A. Pozio, *Decay estimates for partial functional-differential equations*, Nonlinear Analysis **6** (1982), 1253–1266.

M. H. Protter and H. F. Weinberger, *Maximum Principles in Differential Equations*, Prentice-Hall, Englewood Cliffs, NJ, 1967.

C. Robinson, *Stability theorems and hyperbolicity in dynamical systems*, Rocky Mountain J. Math. **7** (1977), 425–437.

W. Rudin, *Real and Complex Analysis*, McGraw-Hill, New York, 1966.

S. H. Saperstone, *Semidynamical systems in infinite dimensional spaces*, Applied Math. Sciences **37** (1981), Springer-Verlag, Berlin.

D. Sattinger, *Monotone Methods in nonlinear elliptic and parabolic boundary value problems*, Indiana Univ. Math. J. **21** (1972), 979–1000.

G. Seifert, *Positively invariant closed sets for systems of delay differential equations*, J. Diff. Eqns. **22** (1976), 292–304.

J. Selgrade, *Asymptotic behavior of solutions to single loop positive feedback systems*, J. Diff. Eqn. **38** (1980), 80–103.

_____, *A Hopf bifurcation in single loop positive feedback systems*, Quart. Appl. Math. (1982), 347–351.

_____, *On the existence and uniqueness of connecting orbits*, Nonlinear Analysis **7** (1983), 1123–1125.

J. Smoller, *Shock Waves and Reaction Diffusion Equations*, Springer-Verlag, New York, 1983.

S. Smale, *On the differential equations of species in competition*, J. Math. Biol. **3** (1976), 5–7.

H. L. Smith, *Cooperative systems of differential equations with concave nonlinearities*, Nonlinear Analysis **10** (1986a), 1037–1052.

_____, *Invariant curves for mappings*, SIAM J. Math. Anal. **17** (1986b), 1053–1067.

_____, *On the asymptotic behavior of a class of deterministic models of cooperating species*, SIAM J. Appl. Math. **46** (1986c), 368–375.

_____, *Competing subcommunities of mutualists and a generalized Kamke theorem*, SIAM J. Appl. Math. **46** (1986d), 856–874.

_____, *Periodic orbits of competitive and cooperative systems*, J. Diff. Eqns. **65** (1986e), 361–373.

_____, *Periodic competitive differential equations and the discrete dynamics of competitive maps*, J. Diff. Eqns. **64** (1986f), 165–194.

_____, *Periodic solutions of periodic competitive and cooperative systems*, Siam J. Math. Anal. **17** (1986g), 1289–1318.

_____, *Monotone semiflows generated by functional differential equations*, J.Diff. Eqns. **66** (1987a), 420–442.

_____, *Oscillation and multiple steady states in a cyclic gene model with repression*, J. Math. Biol. **25** (1987b), 169–190.

———, *Systems of ordinary differential equations which generate an order preserving flow. A survey of results*, SIAM Review **30** (1988), 87–113.

H. L. Smith, B. Tang, and P. Waltman, *Competition in an n-vessel gradostat*, SIAM J. Appl. Math **5** (1991), 1451–1471.

H. L. Smith and B. Tang, *Competition in the gradostat: the role of the communication rate*, J. Math. Bio. **27** (1989), 139–165.

H. L. Smith and H. Thieme, *Convergence for strongly ordered preserving semiflows*, SIAM J. Math. Anal. **22** (1991a), 1081–1101.

———, *Quasi Convergence for strongly ordered preserving semiflows*, SIAM J. Math. Anal. **21** (1990a), 673-692.

———, *Monotone semiflows in scalar non-quasi-monotone functional differential equations*, J. Math. Anal. Appl. **150** (1990b), 289–306.

———, *Strongly order preserving semiflows generated by functional differential equations*, J.Diff. Eqns. **93** (1991b), 332–363.

H. Smith and P. Waltman, *A classification theorem for three dimensional competitive systems*, J. Diff. Eq. **70** (1987), 325–332.

———, *The gradostat: a model of competition along a nutrient gradient*, Microbial Ecol. **22** (1991), 207–226.

———, *Competition for a single limiting resource in continuous culture: the variable-yield model*, SIAM J. Applied Math. **54** (1994), 1113–1131.

——— H. Smith and P. Waltman, P., *The Theory of the Chemostat*, Cambridge Univ. Press, 1995 (to appear).

J. So and P. Waltman, *A nonlinear boundary value problem arising from competition in the chemostat*, Applied Math. and Comp. **32** (1989), 169–183.

H. B. Stewart, *Generation of analytic semigroups by strongly elliptic operators under general boundary conditions*, Trans. Amer. Math. Soc. **259** (1980), 299–310.

P. Takáč, *Asymptotic behavior of discrete-time semigroups of sublinear, strongly increasing mappings with applications in biology*, Nonlinear Analysis **14** (1990), 35–42.

———, *Domains of attraction of generic ω-limit sets for strongly monotone semiflows*, Zeitschrift für Analysis und ihre Anwendungen **10** (1991), 275–317.

Y. Takeuchi, *Diffusion-mediated persistence in two-species competition Lotka-Volterra model*, Math. Biosciences **95** (1989), 65–83.

Y. Takeuchi and Z. Lu, *Permanence and Global stability for competitive Lotka-Volterra diffusion systems*, preprint (1994).

H. R. Thieme, *Convergence results and a Poincaré-Bendixson trichotomy for asymptotically autonomous differential equations*, J. Math. Biol. **30** (1992), 755–763.

C. Travis and W. Post, *Dynamics and comparative statics of mutualistic communities*, J. Theor. Biology **78** (1979), 553-571.

M. P. Vischnevskiĭ, *Stabilization of solutions of weakly coupled cooperative parabolic systems*, (Russian), Mat. Sb. **183** (1992), 45–62.

W. Walter, *Differential and Integral Inequalities*, Springer-Verlag, Berlin, 1970.

J. Wu and H. Freedman, *Monotone semiflows generated by neutral functional differential equations with application to compartmental systems*, Canad. J. Math. **43** (1991), 1098–1120.

J. Wu, *Global dynamics of strongly monotone retarded equations with infinite delay*, J. Integral Eqns. and Appl. **4** (1992), 273–307.

M. Zeeman, *Hopf bifurcations in competitive three-dimensional Lotka-Volterra systems*, Dynamics and Stability of Systems **8 Number 4** (1993).

E. Zeidler, *Nonlinear Functional Analysis and its Applications I Fixed Point Theorems*, Springer-Verlag, New York, 1986.

L. Zhou and C. V. Pao, *Asymptotic behavior of a competition-diffusion system in population dynamics*, Nonlinear Analysis **6** (1982), 1163–1184.

H.-R. Zhu, *The existence of stable periodic orbits for systems of three dimensional differential equations that are competitive*, Ph.D. Thesis, Arizona State University (1991).

H.-R. Zhu and H. Smith, *Stable periodic orbits for a class of three dimensional competitive systems*, J. Diff. Eqn. **110** (1994), 143–156.

Subject Index

A

absorption principle, 8
algebraic multiplicity one, 20, 60, 92, 137
alpha limit set, 32
approximate from below (above), 8
asymptotically stable from above, from below, 15
asymptotically stable point, 15
attracting periodic orbit, 4

B

basin of attraction, 19
Belousov-Zhabotinski reaction, 50
Biochemical control circuit, 58, 93, 139
Borel measure, 79, 86, 105
Brouwer Fixed Point Theorem, 45, 96

C

chain recurrent set, 40, 163
conditionally completely continuous operator, 9
completely continuous operator, 9
Colimiting principle, 6
compact operator, 19, 121
competitive system, 31, 34, 48
concave operator, 153
contracting rectangle, 84, 97
Convergence criterion, 3
convergent point, 2
cooperative system of ODEs, 31, 34, 48
cooperative system of FDEs, 88
cooperative system of PDEs, 129
cooperative and irreducible system of ODEs, 56
cooperative and irreducible system of FDEs, 88
cooperative and irreducible system of PDEs, 133

D

dissipative system, 59

E

equilibrium point, 2
eventually strongly monotone semiflow, 3, 85, 89
essential infimum, 107
exponential ordering, 102

F

falling interval, 37
Field-Noyes model, 50
fixed point index, 17
Floquet multiplier, 64
flow, 32
full orbit of Φ, 26, 62

H

Hausdorff metric, 16, 164
hyperbolic point, 43

I

invariant set, 2
Intersection principle, 6
irreducible matrix, 43, 56

K

Kamke condition, 32
Krein-Rutman Theorem, 19, 20, 93

L

Limit set dichotomy, 5, 8, 21
Limit set separation principle, 8
Lipschitz vector field, 38
Lotka-Volterra system, 35, 58, 64, 69, 94

M

maximum principle, 123
mild solution, 122
monotone semiflow, 2

N

Nagumo condition, 127
negative orbit, 32

Subject Index

N

Nonordering of limit sets, 5
normal cone, 15

O

orbit, 2
order convex, 18
order interval, 15
Order interval trichotomy, 17
order stable equilibrium, 73
omega limit set, 2, 26

P

p-convex, 33
p_m-convex, 48
partial order relation, 1, 48
Perron-Frobenius Theorem, 43, 60
Poincaré-Bendixson Theorem, 40, 41
point spectrum, 92
positive cone, 1, 48
positively invariant set, 2, 34, 81
positive orbit, 32
positive semigroup, 99
principal eigenvalue, 137
principal eigenvector, 137

Q

quasiconvergent point, 2
quasipositive matrix, 60
quasimonotone condition, 78, 129

R

rising interval, 37

ISBN 0-8218-0393-X

S

semiflow, 2
semigroup, 121
Sequential limit set trichotomy, 9
sign-stable matrix, 49
sign-symmetric matrix, 49
Smale's construction, 69
spectral radius, 20, 60
stability modulus, 60, 91, 110
stable point, 15
stable manifold, 43
strictly monotone semiflow, 24
Strong parabolic maximum principle, 124
strongly monotone semiflow, 3
strongly order preserving semiflow, 2
strongly positive operator, 19
sub-equilibrium (strict), 24, 35
super-equilibrium (strict), 24, 35
sub-(super-)solution, 130

T

topologically equivalent vector fields, 38
totally ordered set, 12
type K vector field, 32

U

uniformly elliptic differential operator, 121
uniformly parabolic differential operator, 123
unrelated points, 34

V

vector field, 32

W

w-norm, 21